Mark A. McKibben
Goucher College

Student Solutions Manual
To Accompany

Statistics:
Principles and Methods
Fifth Edition

Richard A. Johnson
University of Wisconsin at Madison

Gouri K. Bhattacharyya
University of Wisconsin at Madison

With Assistance From:
Paul R. Lorczek

WILEY
JOHN WILEY & SONS, INC.

To order books or for customer service please, call 1-800-CALL WILEY (225-5945).

ISBN-13   978- 0-471-71884-0
ISBN-10   0-471-71884-X

Printed in the United States of America

10 9 8 7 6 5 4 3 2 1

Printed and bound by Bind-Rite Graphics

Table Contents

Chapter 1

INTRODUCTION

1.1 The *population* consists of the entire set of responses from all teenagers, 13 to 17 years old, in the United States while the *sample* consists of the responses of the particular 1055 teens contacted in the telephone survey.

1.3 (a) A person living in Chicago is the *unit*.

(b) The *variable of interest* is the characterization of a Chicago resident as an illegal alien or not.

(c) The *statistical population* consists of the collection of illegal alien characterizations of each resident of the city of Chicago.

1.5 (a) An undergraduate student is the *unit*.

(b) The *statistical population* consists of the collection of ownership/non-ownership characteristics for all undergraduate students at the school. The *sample* consists of the set of ownership/non-ownership characteristics for each of the ten joggers.

(c) The sample is likely to be non-representative because joggers might strongly prefer to exercise with a small size MP3 player rather than larger alternative music sources. They would then be among the first to purchase.

1.7 The newspaper is suggesting that the *population* is the collection of preferences for each adult in the city while the *sample* is the collection of preferences of the particular persons who sent in their votes. This sample is apt to be non-representative because those persons in the sample are self-selected. Only the few who feel very strongly positive will likely send in a vote.

1.9   (a) This is anecdotal. No data given.

   (b) The yes/no answer regarding multiple credit cards, for each of the 22 students, is the sample on which the statement is based.

   (c) The yes/no answer regarding destination outside the continental United States, for each of the 55 people at the airport, is the sample on which the statement is based.

1.11 The term "too long" is not well defined. By asking a number of people, we may determine that 5 minutes is too long. Further, the time will not be the same for all people. One improved statement of the purpose is:

   Purpose: Determine if over half the persons take over 5 minutes to get cash during the lunch hour.

1.13 First number the classrooms 1 to 35. In Table 1, we started in row 20 using columns 29 and 30. Reading downward, and ignoring 00 and numbers above 35, we selected rooms 8, 7, 1, and 9.

1.15 We started in row 10 and read down column 9 and then down column 6 from the top. We ignored the second digit in a pair, and kept reading, where the two digits were the same. That type of assignment of students is not allowed.

20 pairs of random digits

4,0	1,2	4,2	5,2	2,1
5,1	3,2	2,0	7,6	5,4
0,1	2,5	2,7	3,6	5,4
2,4	5,7	3,5	6,7	7,2

   (a) $4/20 = .20$

   (b) $9/20 = .45$

   (c) $7/20 = .35$

1.17

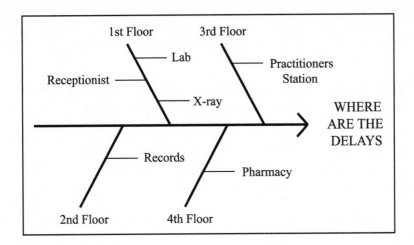

1.19 (a) The miniature poodles could never be observed even if they greatly outnumber the Great Danes. Only the big dogs can volunteer to show they were inside the fence.

(b) Persons who call-in their opinions are self-selected because they have strong opinions. This is analogous to the big dogs who are the only volunteers to show they were inside the fence.

Chapter 2

ORGANIZATION AND DESCRIPTION OF DATA

2.1   (a) The percentage in other classes is $100 - 38.6 - 12.8 - 10.1 - 9.9 - 7.7 = 20.9$ %

     (b)

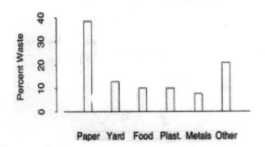

  (c) The percentage of waste that is paper or paperboard $= 38.6$ %

     The percentage of waste in the top two categories is $38.6 + 12.8 = 51.4$

     The percentage in the top five categories is $38.6 + 12.8 + 10.1 + 9.9 + 7.7 = 79.1$

2.3   The frequency table for blood type is

Blood type	Frequency	Relative Frequency
O	16	$0.40 = 16/40$
A	18	$0.45 = 18/40$
B	4	$0.10 = 4/40$
AB	2	$0.05 = 2/40$
Total	40	1.00

2.5   (a) The table of relative frequencies for workers in the department is

Mode of Transportation	Frequency	Relative Frequency
Drive alone	25	$0.625 = 25/40$
Car pool	3	$0.075 = 3/40$
Ride bus	7	$0.175 = 7/40$
Other	5	$0.125 = 5/40$
Total	40	1.000

(b) The pie chart for workers in the department is

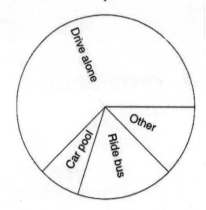

2.7   There are overlapping classes in the grouping. A report of 3 stolen bicycles will fall in two classes.

2.9   There is a gap. The response 5 close friends does not fall in any class. The last class should be 5 or more.

2.11 (a) Yes. (b) Yes. (c) Yes. (d) No. (e) No.

2.13 (a) The relative frequencies are $9/50 = .18$, .48, .26, and .08 for 0, 1, 2, and 3 bags, respectively.

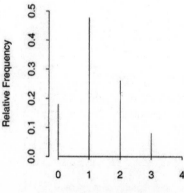

Number of bags checked

(b) Almost half of the passengers check exactly one bag. The longest tail is to the right.

(c) The proportion of passengers who fail to check a bag is $9/50 = .18$.

2.15 The dot diagram of amounts of radiation leakage is

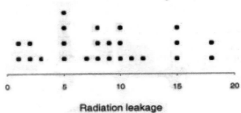

2.17 (a) The dot diagram of number of CFUs is

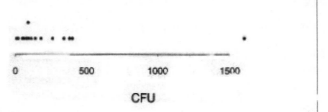

(b) There is a long tail to the right with one extremely large value of 1600 CFU units.

(c) There is one day so the proportion is $1/15 = 0.067$

2.19 (a) In the following frequency distribution of lizard speed (in meters per second), the left endpoint is included in the class interval but not the right endpoint.

Class Interval	Frequency	Relative Frequency
.45 to .90	2	0.067
.90 to 1.35	6	0.200
1.35 to 1.80	11	0.367
1.80 to 2.25	5	0.167
2.25 to 2.70	6	0.200
Total	30	1.001 (rounding error)

(b) All of the class intervals are of length 0.45 so we can graph rectangles whose heights are the relative frequency. The histogram is

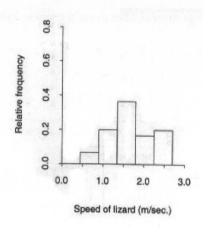

2.21   The frequencies are 5, 11, 13, 7, 8 and 5, respectively. The class intervals have

unequal widths so the rectangles have: $\text{height} = \dfrac{\text{relative frequency}}{\text{width of interval}}$

For the interval 5.2 to 5.6, height $= (5/49)/.4 = 0.0255$.

For the interval 7.2 to 8.4, height $= (5/49)/1.2 = 0.0850$.

The histogram has a long right tail.

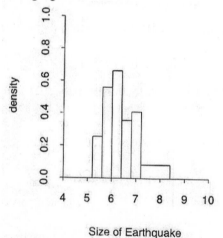

2.23   The stem-and-leaf display of the amount of iron present in the oil is

$$
\begin{array}{r|l}
0 & 6 \\
1 & 2234455567777889 \\
2 & 000000222445567799 \\
3 & 022444566 \\
4 & 1167 \\
5 & 12 \\
\end{array}
$$

2.25 The double-stem display of the amount of iron present in the oil is

```
0 | 6
1 | 22344
1 | 55567777889
2 | 00000022244
2 | 5567799
3 | 022444
3 | 566
4 | 11
4 | 67
5 | 12
```

2.27 The five-stem display of the Consumer Price Index in 2001 for the given cities is

```
15 | 5
15 |
15 | 9
16 |
16 |
16 |
16 | 7
16 | 8
17 | 0
17 | 22333
17 | 4
17 | 677
17 | 88
18 | 11
18 | 2
18 |
18 | 67
18 |
19 | 011
```

2.29 (a) The median is 3. The sample mean is

$$\bar{x} = \frac{2+5+1+4+3}{5} = \frac{15}{5} = 3$$

(b) The mean is

$$\bar{x} = \frac{26+30+38+32+26+31}{6} = \frac{183}{6} = 30.5$$

The ordered measurements are: 26, 26, 30, 31, 32, 38

$$\text{median} = \frac{30+31}{2} = 30.5$$

(c) The sample mean is

$$\bar{x} = \frac{-1+2+0+1+4-1+2}{7} = 1$$

The ordered measurements are: $-1, -1, 0, 1, 2, 2, 4$.
The median is 1.

2.31 (a) $\bar{x} = 3810/15 = 254$.

(b) The ordered observations are:

10	20	50	60	80	90		90	110
140	180	260	340	380	400	1600		

So, the median is 110 CFU units. The one very large observation makes the sample mean much larger. Hence, the sample median is better to use in this instance.

2.33 The mean is $956/12 = 79.67$. The claim ignores variability and is not true. It is certainly unpleasant with a daily maximum temperature $105°F$ in July.

2.35 (a) $\bar{x} = 212/25 = 8.48$
(b) The sample median is 8. Since the sample mean and median are about the same, either of them can be used as an indication of radiation leakage.

2.37 The mean, 118.05, is one measure of center tendency and the median, 117.00, is another. The value 118.05 tells us that, on average, that a baby weighed 118.05 ounces. The median tells us that about half of the babies weighed at least 117 ounces while roughly half weighed at most 117 ounces.

2.39 $\bar{x} = \dfrac{429+425+471+422+432+444+454}{7} = \dfrac{3077}{7} = 439.6$

2.41 (a) $\bar{x} = 271/40 = 6.775$ days.
   (b) Sample median $= (6+7)/2 = 6.5$. Both the sample mean and the sample median give a good indication of the amount of mineral lost.

2.43 Sample median $= (176+187)/2 = 181.5$ (minutes).

2.45 (a) The dot diagram for the diameters (in feet) of the Indian mounds in southern Wisconsin is

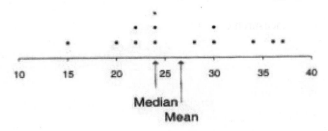

   (b) $\bar{x} = 346/13 = 26.62$. Sample median $= 24$.
   (c) $13/4 = 3.25$, so we count in 4 observations. $Q_1 = 22$ and $Q_3 = 30$.

2.47 (a) Median $= (152+154)/2 = 153$.
   (b) $40/4 = 10$, so we need to count in 10 observations. The 11-th smallest observation also satisfies the definition.

$$Q_1 = \frac{135+136}{2} = 135.5, \quad Q_3 = \frac{166+167}{2} = 166.5$$

2.49 The ordered data are

   50   57   68   69   72   73   73   80   82   91
   92   93   94   96   96   100   102   104   105   106
   108   109   118   118   127

   Since the number of observations is 25, the median or second quartile is the 13th ordered observation in the list. The first quartile is the 7th observation.

$$Q_1 = 73 \quad Q_2 = 94 \quad Q_3 = 105$$

2.51 (a) The ordered observations are

   10   20   50   60   80   90   90   110
   140   180   260   340   380   400   1600

Since the sample size is 15, the median is the 8th observation 110. To obtain Q_1, we find $15/4 = 3.75$ so the first quartile is the 4th observation in the ordered list.

$$Q_1 = 60 \quad Q_3 = 340$$

(b) The 90th percentile requires us to count in at least $.9(15) = 13.5$ or 14 observations. The 90th sample percentile $= 400$.

2.53 (a) The ordered data are 73, 74, 76, 76, 80. The median is $76°F$ and the mean is $\bar{x} = 379/5 = 75.8°\,F$.

(b) The mean of $(°F - 32)$ is $\bar{x} - 32$ by property (i) of Exercise 2.52 with $c = -32$. By property (ii)

$$\text{mean of } \frac{5}{9}(°F - 32) = \frac{5}{9}(\text{mean of } (°F - 32))$$

$$= \frac{5}{9}(\bar{x} - 32) = \frac{5}{9}(75.8 - 32) = 24.33°\,C$$

By similar properties for the median

$$\text{median of } \frac{5}{9}(°F - 32) = \frac{5}{9}(\text{median of } (°F) - 32) = \frac{5}{9}(76 - 32) = 24.44°\,C$$

2.55 (a)

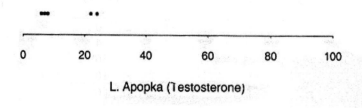

L. Apopka (Testosterone)

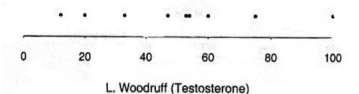

L. Woodruff (Testosterone)

(b) Lake Apopka $\bar{x} = \dfrac{67}{5} = 13.40$ Lake Woodruff $\bar{x} = \dfrac{454}{9} = 50.44$

(c) From the dot diagrams, the males in Lake Apopka have lower levels of testosterone and their sample mean is only about one-third of that for males in (un-contaminated) Lake Woodruff. This finding is consistent with the environmentalists' concern that the contamination has affected the testosterone levels and the reproductive abilities.

2.57 (a) We carry out all necessary calculations in the following table. The mean is $\bar{x} = 15/3 = 5$.

	x	$x - \bar{x}$	$(x - \bar{x})^2$
	8	3	9
	3	−2	4
	4	−1	1
Total	15	0.0	14

(b) The variance and the standard deviation are
$$s^2 = \frac{14}{3-1} = 7 \ , \ s = \sqrt{7} = 2.646$$

2.59 (a) We carry out all necessary calculations in the following table. The mean is $\bar{x} = 32/4 = 8$.

	x	$x - \bar{x}$	$(x - \bar{x})^2$
	8	0	0
	6	−2	4
	14	6	36
	4	−4	16
Total	32	0.0	56

(b) The variance and the standard deviation are

$$s^2 = \frac{56}{4-1} = 18.667 \ , \ s = \sqrt{18.667} = 4.320$$

2.61 We carry out all necessary calculations in the following table.

	x	x^2
	8	64
	3	9
	4	16
Total	15	89

The variance is

$$s^2 = \frac{1}{n-1}\left(\sum x^2 - \frac{\left(\sum x\right)^2}{n}\right) = \frac{1}{2}\left(89 - \frac{15^2}{3}\right) = \frac{1}{2}(89-75) = 7$$

2.63 (a) $s^2 = (34 - 12^2/5)/4 = 1.30$.

   (b) $s^2 = (19 - (-7)^2/6)/5 = 2.167$.

   (c) $s^2 = (499 - 59^2/7)/6 = 0.286$.

2.65 $s = \sqrt{(9726 - 346^2/13)/12} = 6.5643$.

2.67 (a) $s^2 = (3,140,900 - 3810^2/15)/14 = 155,225.7$.

   (b) $s = \sqrt{155225.7} = 393.99$.

   (c) $s^2 = (580900 - 2210^2)/14)/13 = 17,848.9$. so $s = \sqrt{17848.9} = 133.6$. The single very large value greatly inflates the standard deviation.

2.69 (a) $\bar{x} = 1862/10 = 186.2$.

   (b) $s^2 = (353796 - 1862^2/10)/9 = 787.96$.

   (c) $s = \sqrt{787.96} = 28.07$

2.71 (a) Median $= 68.4$.

   (b) $\bar{x} = 478.4/7 = 68.343$.

   (c) $s^2 = (32730.34 - 478.4^2/7)/6 = 5.853$. Hence $s = 2.419$.

2.73 (a) The measure of variation displayed is 15.47, the sample standard deviation. The sample variance is $s^2 = 15.47^2 = 239.321$.

   (b) The interquartile range is $Q_3 - Q_1 = 131.00 - 106.00 = 25.00$. This means the center half of the data span an interval of length 25 ounces.

   (c) Any value greater than 15.47 would correspond to greater variation.

2.75 In Exercise 2.47, we determined that $Q_1 = 135.5$ and $Q_3 = 166.5$. Hence,

   Interquartile range $= Q_3 - Q_1 = 166.5 - 135.5 = 31.0$ points.

2.77 No. Typically, the middle half of a data set is much more concentrated than the sum of the two quarters, one in each tail. As an example, for the water quality data of Exercise 2.17, the range is $1600 - 10 = 1590$ because of one extremely large observation. From the quartiles determined in Exercise 2.51, the interquartile range is $340 - 60 = 280$. The range is six times larger than the interquartile range.

2.79 (a) $\bar{x} = 6.775$ and $s = \sqrt{19.4096} = 4.406$.

(b) The proportion of the observations are given in the following table:

	$\bar{x} \pm s$	$\bar{x} \pm 2s$	$\bar{x} \pm 3s$
Interval:	(2.369, 11.181)	(−2.037, 15.587)	(−6.443, 19.993)
Proportion:	26/40 = 0.65	38/40 = 0.95	40/40 = 1.00
Guidelines:	0.68	0.95	0.997

(c) We observe a good agreement with the proportions suggested by the empirical guideline.

2.81 (a) $\bar{x} = 25.160$ and $s = \sqrt{114.790} = 10.714$.

(b) The proportion of the observations are given in the following table:

	$\bar{x} \pm s$	$\bar{x} \pm 2s$	$\bar{x} \pm 3s$
Interval:	(14.446, 35.874)	(3.732, 46.588)	(−6.982, 57.302)
Proportion:	36/50 = 0.72	47/50 = 0.94	50/50 = 1.00
Guidelines:	0.68	0.95	0.997

(c) We observe a good agreement with the proportions suggested by the empirical guideline.

2.83 (a) $z = \dfrac{102 - 118.05}{15.47} = -1.037$

(b) $z = \dfrac{144 - 118.05}{15.47} = 1.677$

2.85 For males, the minimum and the maximum horizontal velocity of a thrown ball are 25.2 and 59.9 respectively. The quartiles are:

$$Q_1 = (38.6 + 39.1)/2 = 38.85,$$
$$\text{median} = (45.8 + 48.3)/2 = 47.05,$$
$$Q_3 = (49.9 + 51.7)/2 = 50.8.$$

For females, the minimum and the maximum horizontal velocity of a thrown ball are 19.4 and 53.7 respectively. The quartiles are

$$Q_1 = 25.7, \text{ median} = 30.3, Q_3 = 33.5.$$

The boxplot of the male and female throwing speed are

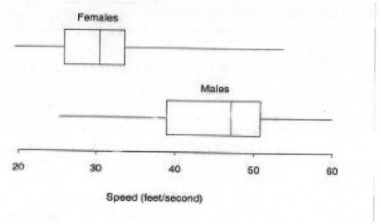

Comparing the two boxplots, we can see that males throw the ball faster than females.

2.87 (a)

$$\bar{x} = \frac{894.0}{24} = 37.25 \quad \text{and} \quad s = \sqrt{\frac{34266 - (894)^2 / 24}{24 - 1}} = 6.476$$

(b) Since $\bar{x} - 2s = 37.25 - 2(6.476) = 24.3$ and $\bar{x} + 2s = 37.25 + 2(6.476) = 50.2$ only the increase of 23 for Honolulu and 51 for Denver lie outside the interval. The proportion $22 / 24 = 0.92$ lie within the interval.

2.89 (a)

$$\bar{x} = \frac{-141}{16} = -8.813 \quad \text{and} \quad s = \sqrt{\frac{7387 - (-141)^2 / 16}{16 - 1}} = 20.239$$

(b) The very large increase, 67 days for Sacramento, shifts the mean by almost 4 days. If Sacramento is excluded from the calculation the sample mean increases to $-74 / 15 = -4.93$ days.

2.91 (a) The ordered data are

$$-5 \quad 4 \quad 5 \quad 8 \quad 11 \quad 12 \quad 16 \quad 26 \quad 43 \quad 47 \quad 52 \quad 55$$

median $= (12 + 16) / 2 = 14$ seats lost.

(b) The maximum number of seats lost, 55, occurred when Harry S. Truman was President. The minimum number, -5 or a gain, occurred during W. Clinton's second term as President.

(c) range $= 55 - (-5) = 60$

2.93

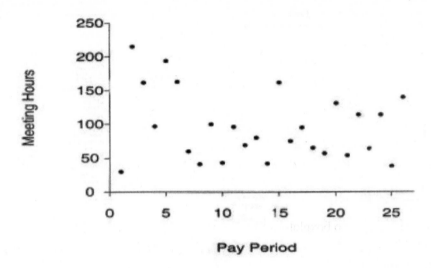

The value 215 from the second pay period looks high and 194 from the fifth period is possibly high.

2.95 We calculate $\bar{x} = 2501/26 = 96.2$ and $s = \sqrt{65254/25} = 51.1$ so the upper limit is $\bar{x} + 2s = 198.4$ and the lower limit is $\bar{x} - 2s = -6.0$ which we take as 0.

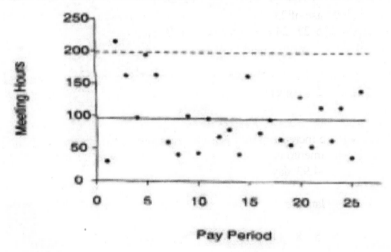

Only the value 215 from the second pay period is out of control.

2.97

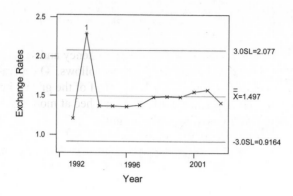

Control Chart for Exchange Rates

The process appears to be in statistical control. Only the value of 2.29 in 1993 is out of control.

2.99 (a) The relative frequencies of the occupation groups are:

	Relative Frequency	
	1980	2001
Goods Producing	0.284	0.190
Service (Private)	0.537	0.652
Government	0.179	0.158
Total	1.000	1.000

(b) The proportion of persons in private service occupations has increased while the proportions in goods producing and government have decreased from 1980 to 2001.

2.101   The dot diagrams of heights for the male and female students are

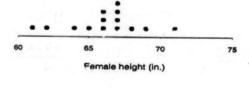

Female height (in.)

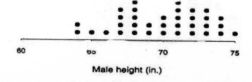

Male height (in.)

2.103 (a) Yes. The exact number of lunches is the sum of the frequencies of the first three classes.
   (b) Yes. The exact number of lunches is the sum of the frequencies of the last three classes.
   (c) No.

2.105 (a) The mean, 227.4, is one measure of center tendency and the median, 232.5, is another. These values may be interpreted as follows. On average, the 20 grizzly bears weigh 227.4 pounds apiece. Half of the grizzly bears sampled weighed at least 232.5 pounds while half weighed at most 232.5 pounds.
   (b) The sample standard deviation is 82.7 pounds.
   (c) The z score for a weight of 320 pounds is
$$z = \frac{320 - 227.4}{82.7} = 1.12$$

2.107 (a) Sample median $= (9+9)/2 = 9$.
   (b) $\bar{x} = 271/30 = 9.033$.
   (c) The sample variance is
$$s^2 = \frac{1}{29}\left(2561 - \frac{271^2}{30}\right) = 3.895.$$

2.109 (a) $\bar{x} = 7$, $s = 2$
   (b) By the properties, the new data set $x + 100$ has sample mean $= (7 + 100)$ $= 107$ and standard deviation 2. By direct calculation, we verify
$$\frac{106 + 108 + 104 + 109 + 108}{5} = 107$$
$$\frac{(106-107)^2 + (108-107)^2 + (104-107)^2 + (109-107)^2 + (108-107)^2}{4} = 4$$
   (c) By the properties, the new data set $-3x$ has sample mean $= -3(7) = -21$ and standard deviation $|-3|s = 3(2) = 6$. By direct calculation, we verify
$$\frac{-18 - 24 - 12 - 27 - 24}{5} = -21$$
$$\frac{(3)^2 + (-3)^2 + (9)^2 + (-6)^2 + (-3)^2}{4} = 9\left(\frac{1+1+9+4+1}{4}\right) = 9s^2$$

2.111 (a) The dot diagrams are

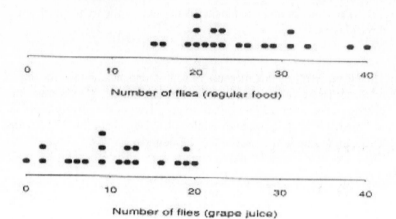

(b) From the dot diagrams we can see the number of flies (grape juice) is centered at about 11 and the number of flies (regular food) is centered near 25. The spread looks about the same.

(c) Regular food: $\bar{x} = 25.1$, $s = 6.84$.
    Grape juice: $\bar{x} = 10.6$, $s = 6.07$.

2.113 (a) $\bar{x} = 5.38$ and $s = 3.42$.
       (b) Median $= 5$.
       (c) Range $= 13 - 0 = 13$.

2.115 (a) and (b). We have $\bar{x} = 4643/54 = 85.981$ and
       $$s = \sqrt{\left[824971 - (4643)^2/54\right]/53} = 89.628.$$

	$\bar{x} \pm s$	$\bar{x} \pm 2s$	$\bar{x} \pm 3s$
Interval:	$(-3.647, 175.609)$	$(-93.275, 265.237)$	$(-182.903, 354.865)$
Proportion:	$49/54 = 0.907$	$51/54 = 0.944$	$52/54 = 0.963$
Guidelines:	0.68	0.95	0.997

(c) The count 49 is a high. The histogram shows a long right-hand tail. The shorter left-hand tail accounts for the high count in the first interval.

2.117 (a) Median $= 4.505$, $Q_1 = 4.30$ and $Q_3 = 4.70$.

       (b) 90th percentile $= = (4.80 + 5.07)/2 = 4.935$.

       (c) $\bar{x} = 4.5074$ and $s = 0.368$.

(d) The boxplot of acid rain in Wisconsin is

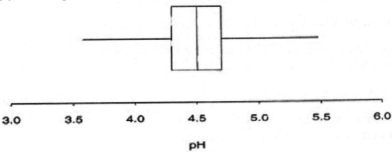

2.119 (a) Median $= 6.3$, $Q_1 = 5.9$ and $Q_3 = 6.9$.

(b) $\bar{x} = 314.8/49 = 6.424$ and $s = \sqrt{\left[2047.4 - (314.8)^2/49\right]/48} = .721$.

(c)

Class Interval (%)	Frequency	Relative Frequency
(5.2, 5.6]	5	0.1020
(5.6, 6.0]	11	0.2245
(6.0, 6.4]	13	0.2653
(6.4, 6.8]	7	0.1429
(6.8, 7.2]	8	0.1633
(7.2, 8.4]	5	0.1020
Total	49	1.0000

We use the convention that the right endpoint is included in the class interval.

(d) The boxplot of the data is

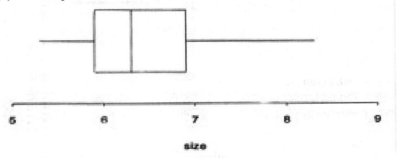

2.121 (a)

Winning Times in Minutes and Seconds

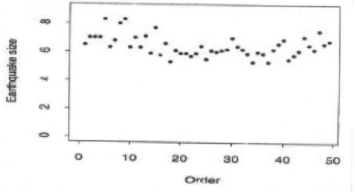

(b) It is not reasonable, because a frequency distribution would not show the systematic decrease of the winning times over the years, which is the main feature of these observations.

2.123 (a) The time series plot is given below

(b) There seems to be a systematic decrease and then increase in level.

2.125 (a) The partial MINITAB output is

Variable	N	Mean	Median	StDev
Speed	30	1.724	1.665	0.573

Variable	Minimum	Maximum	Q1	Q3
Speed	0.500	2.670	1.288	2.125

(b) The partial MINITAB output for the acid rain data.

	N	MEAN	MEDIAN	STDEV
PH	50	4.5074	4.5050	0.3681

	MIN	MAX	Q1	Q3
PH	3.5800	5.4800	4.2950	4.7000

(Note that MINITAB uses a slightly different convention for determining Q_1 and Q_3.)

2.127 The partial MINITAB output for the data set in Table 4.

Variable	N	Mean	Median	StDev
Booksal	40	306.7	301.7	143.3

Variable	Minimum	Maximum	Q1	Q3
Booksal	16.0	621.4	217.5	426.4

2.129 The mean and standard deviation given by MINITAB are the rounded off values of the answer given by SAS.

2.131 (a) The histogram of the alligator data is

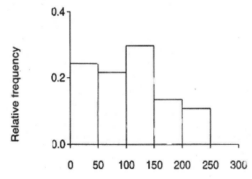

(b)

$$\bar{x} = \frac{4035}{37} = 109.1 \quad s = \sqrt{\frac{155672}{37-1}} = 65.8$$

2.133 (a)

Descriptive Statistics: Maltextr

Variable	N	Mean	Median	StDev
Maltextr	40	77.458	77.400	1.101

(b) The ordered observations are

75.3 75.7 75.9 75.9 76.2 76.3 76.4 76.4 76.6 76.6
76.7 76.9 76.9 77.0 77.0 77.1 77.4 77.4 77.4 77.4
77.4 77.5 77.6 77.6 77.8 77.9 77.9 77.9 77.9 77.9
78.0 78.1 78.3 78.4 78.4 78.5 79.1 79.2 80.0 80.4

There are 40 observations so the median $= (77.4+77.4)/2 = 77.4$. The first quartile is the average of the $40/4 = $ 10th and 11th observations in the sorted list. $Q_1 = (76.6+76.7)/2 = 76.65$ and $Q_1 = (77.9+78.0)/2 = 77.95$.

(c) The interval $\bar{x} \pm s$ or $(76.36, 78.56)$ has relative frequency $30/40 = .75$ compared to .683. The interval $\bar{x} \pm 2s$ or $(75.26, 79.66)$ has relative frequency $38/40 = .95$ compared to .95. The interval $\bar{x} \pm 3s$ or $(74.16, 80.76)$ has relative frequency 1 compared to .997. The agreement is quite good.

Chapter 3

DESCRIPTIVE STUDY OF BIVARIATE DATA

3.1   (a) The table, with completed marginal totals, is:

| | Degree of Nausea | | | | |
	None	Slight	Moderate	Severe	Total
Pill	43	36	18	3	100
Placebo	19	33	36	12	100
Total	62	69	54	15	200

(b) The relative frequencies, by row, are

| | Degree of Nausea | | | | |
	None	Slight	Moderate	Severe	Total
Pill	0.43	0.36	0.18	0.03	1.00
Placebo	0.19	0.33	0.36	0.12	1.00

(c) A much higher proportion, 0.43, of pill takers avoided nausea as compared to the proportion 0.19 among those who took the placebo. Also the proportion of persons suffering moderate and severe nausea was much lower among those receiving the pill.

3.3   To compare aging, we find the relative frequencies, by row.

Model	≤ 20 years	> 20 years	Total
B7	0.423	0.577	1.000
B27	0.736	0.264	1.000
B37	0.991	0.009	1.000

The proportion of planes over 20 years old is highest for the (oldest) model B7 and it decreases, to a negligible amount for the (newest) model B37.

3.5   (a) The two-way frequency table is:

		Iron		
		Low	High	Total
Alkalinity	Low	8	2	10
	High	4	5	9
Total		12	7	19

(b) The relative frequencies are:

		Iron		
		Low	High	Total
Alkalinity	Low	0.421	0.105	0.526
	High	0.211	0.263	0.474
Total		0.632	0.368	1.000

(c) The relative frequencies, by row, are:

		Iron		
		Low	High	Total
Alkalinity	Low	0.800	0.200	1.000
	High	0.444	0.556	1.000

3.7   (a) The two-way frequency table is:

	Major				
	B	H	P	S	Total
Male	12	4	5	14	35
Female	6	0	4	4	14
Total	18	4	9	18	49

(b) The relative frequencies are:

	Major				
	B	H	P	S	Total
Male	0.245	0.082	0.102	0.286	0.715
Female	0.122	0	0.082	0.082	0.286
Total	0.367	0.082	0.184	0.368	1.001

Alternate solution using Minitab:

When a data set is large, it is useful to enter it once on a computer and instruct it to do the counting. To do so, we encode gender and intended major as numbers. We choose 0 if male, 1 if female and $1 = B$, $2 = H$, $3 = P$, $4 = S$

With the coded data in a file called 2.94.dat.

```
ROWS: GENDER    COLUMNS: MAJOR

              1       2       3       4      ALL

      0      12       4       5      14       35
      1       6       0       4       4       14
    ALL      18       4       9      18       49


CELL CONTENTS --
              COUNT
```

We can also calculate relative frequencies by row. More precisely, $100 \times$ (relative frequency) is obtained from the MINITAB command

        TABLE C1 C4:

        ROWPERCENT.

```
ROWS: GENDER      COLUMNS: MAJOR

            1        2        3        4       ALL
    0     34.29    11.43    14.29    40.00   100.00
    1     42.86      --     28.57    28.57   100.00
  ALL     36.73     8.16    18.37    36.73   100.00

CELL CONTENTS --
              % OF ROW
```

3.9 (a) The proportions are calculated by row $0.548 = 23/42$ and so on.

	Male	Female	Total
English	0.548	0.452	1.000
Computer science	0.844	0.156	1.000

(b) There appears to be gender bias, favoring males, in the Computer Science department. The relative frequencies in the English department do not indicate the obvious presence of bias.

3.11 (a) The proportions, by row, for each condition are

Good Condition			
	Died	Survived	Total
Research hospital	0.021	0.979	1.000
Community hospital	0.027	0.973	1.000

Bad Condition

	Died	Survived	Total
Research hospital	0.050	0.950	1.000
Community hospital	0.070	0.930	1.000

(b) The research hospital has a higher proportion of patients in good condition that survive, 0.979 vs.0.973, and a higher proportion of patients in poor condition that survive, 0.950 vs. 0.930. Whether you are in bad or in good condition, you should prefer the research hospital.

(c) We have reached just the opposite conclusion of that reached in Exercise 3.10. In this example of Simpson's paradox, the condition of the patient acted as the lurking variable. The proportion of patients in poor condition is much higher at the research hospital so that kept down their overall survival rate calculated in Exercise 3.10.

3.13 (a) Of course, the fact that 21 out of 57, or proportion .368 quit smoking, by itself, would seem to be stronger evidence. Intuitively, we tend to think incorrectly that no persons would have quit without the medicated patch.

(b) Most people respond positively when they are given attention. The placebo trials make it possible to treat all subjects alike except for the presence or absence of medication. Twenty percent, 11 out of 55, responded positively to the procedure, even without the medication. This makes the success of the medicated patch less spectacular but provides a proper frame of reference.

3.15 (a) Positive – more sales persons should be able to see more people and sell more real estate.

(b) Positive – in general better players get paid higher salaries.

(c) Positive – one would expect sales to increase with the amount of TV advertising of the cola.

(d) Negative – strength diminishes with age after middle age.

3.17 No. The value of r can be small even is there is a strong relationship along a curve as illustrated in Figure 2 of the text.

3.19 (a) A computer calculation gives $r = -0.460$ for males.
        The scatter plot diagram for males and the multiple scatter plot are shown on the next page.

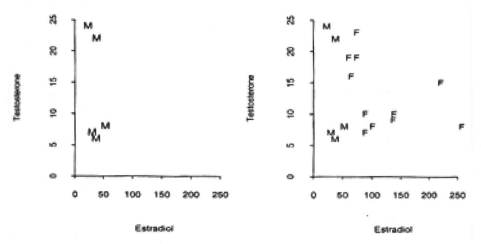

(b) A computer calculation gives $r = -0.415$ for females.

(c) Both have about the same testosterone levels but Females have higher levels of estradiol and are more variable.

3.21 Only Figure 7(c) has a northwest-southeast pattern indicating a negative value for r. Since the tightest pattern, indicating the largest r, is in Figure 7(a) the matches are

(a) $r = -0.3$ and Figure 7(c).   (b) $r = 0.1$ and Figure 7(b).

(c) $r = 0.9$ and Figure 7(a).

3.23 Identifying the sums of squares about the means as S_{xx}, S_{yy}, and S_{xy} respectively, we find

$$r = \frac{S_{xy}}{\sqrt{S_{xx}}\sqrt{S_{yy}}} = \frac{-204.3}{\sqrt{530.7}\sqrt{215.2}} = -0.605$$

3.25 Let $x =$ the amount of hydrogen and $y =$ the amount of carbon. Then, with $n = 11$, we calculate

$$\sum x = 533.80 \quad , \quad \sum y = 621.00$$
$$\sum x^2 = 43{,}124.84 \qquad \sum y^2 = 48{,}624.58 \qquad \sum xy = 43{,}760.84$$

so

$$S_{xx} = \sum x^2 - \frac{\left(\sum x\right)^2}{n} = 43{,}124.84 - \frac{(533.80)^2}{11} = 17{,}220.98$$

$$S_{yy} = \sum y^2 - \frac{\left(\sum y\right)^2}{n} = 48{,}624.58 - \frac{(621.00)^2}{11} = 13{,}566.31$$

$$S_{xy} = \sum xy - \frac{\left(\sum x\right)\left(\sum y\right)}{n} = 43{,}760.84 - \frac{(533.80)(621.00)}{11} = 13{,}625.40$$

Consequently,

$$r = \frac{S_{xy}}{\sqrt{S_{xx}}\sqrt{S_{yy}}} = \frac{13,625.40}{\sqrt{17,220.98}\sqrt{13,566.31}} = 0.891$$

3.27 (a) The scatter diagram is shown in below. The pattern runs from lower left to upper right and is not very tight. We estimate $r = 0.2$.

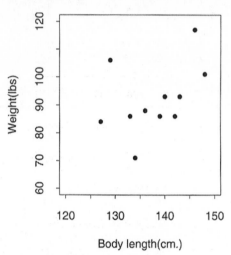

(b)   Let $x =$ length and $y =$ weight. We calculate

$$\sum x = 1511 \quad , \quad \sum y = 1011$$
$$\sum x^2 = 208,153 \quad \sum y^2 = 94,453 \quad \sum xy = 139,141$$

so

$$S_{xx} = \sum x^2 - \frac{(\sum x)^2}{n} = 208,153 - \frac{(1511)^2}{11} = 596.545$$

$$S_{yy} = \sum y^2 - \frac{(\sum y)^2}{n} = 94,453 - \frac{(1011)^2}{11} = 1532.909$$

$$S_{xy} = \sum xy - \frac{(\sum x)(\sum y)}{n} = 139,141 - \frac{(1511)(1011)}{11} = 266.364$$

Consequently,

$$r = \frac{S_{xy}}{\sqrt{S_{xx}}\sqrt{S_{yy}}} = \frac{266.364}{\sqrt{596.545}\sqrt{1532.909}} = 0.279$$

(c) The multiple scatter diagram reveals different patterns and one possible F outlier.

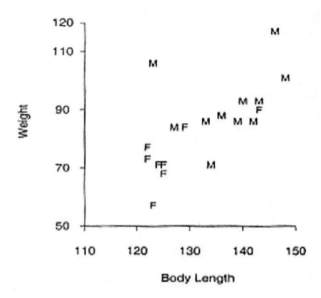

3.29 (a)

Scatter Diagram for Exercise 3.29

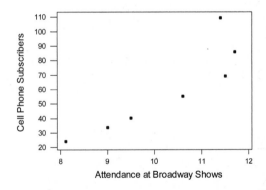

(b) The scatter diagram exhibits a strong correlation but it is hard to imagine any causal relationship between an increase in Broadway show attendance and an increase in the inmate population so we suspect the presence of lurking variables. A steadily increasing population seems the likely culprit.

3.31 (a) The scatter diagram is shown below.

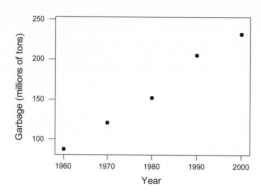

(b) There is a tight southwest to northeast pattern indicating a strong positive correlation. The later years have the largest garbage values and the early years the smallest.

(c) Population size is also increasing and most likely, even with more recycling, more people mean more garbage.

3.33 (a) Let $x = (\text{year} - 1960)$ and $y =$ amount of garbage (mil.tons). Then, with $n = 5$, we calculate

$$\sum x = 100 \quad , \quad \sum y = 798$$
$$\sum x^2 = 3000 \quad \sum y^2 = 141,338 \quad \sum xy = 19,680$$

so

$$S_{xx} = \sum x^2 - \frac{\left(\sum x\right)^2}{n} = 3000 - \frac{(100)^2}{5} = 1000$$

$$S_{yy} = \sum y^2 - \frac{\left(\sum y\right)^2}{n} = 141,338 - \frac{(798)^2}{5} = 13,977.2$$

$$S_{xy} = \sum xy - \frac{\left(\sum x\right)\left(\sum y\right)}{n} = 19,680 - \frac{(100)(798)}{5} = 3720$$

Consequently,

$$r = \frac{S_{xy}}{\sqrt{S_{xx}}\sqrt{S_{yy}}} = \frac{3720}{\sqrt{1000}\sqrt{13,977.2}} = 0.995.$$

(b) The correlation is still 0.995. Since year is a linear transformation of (year – 1960)

$$\text{year} = 1 \cdot (\text{year} - 1960) + 1960$$

The deviation for each year, or $\left(\text{year} - \overline{\text{year}} \right)$ equals the same deviation for $\left(\text{year} - \overline{1960} \right)$ as you may verify. Consequently the sum of squares for years and the sum of cross-products remain the same and the correlation is unchanged (see Exercise 3.30).

3.35 The value of y at $x = 1$ is $2 + 3(1) = 5$ and the value at $x = 4$ is $2 + 3(4) = 14$. The line is shown below. The intercept is 2, the value of y at $x = 0$, and the slope is 3, the coefficient of x.

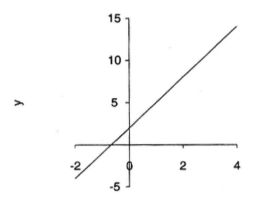

3.37 (a) $y = 10(41) - 155 = 255$.

(b) Note that $y = 0$ if $10x = 155$ or if $x = 15.5$. A profit will be made if 16 or more units are sold.

3.39 (a) (b) The scatter diagram and the visually drawn dotted line are shown below.

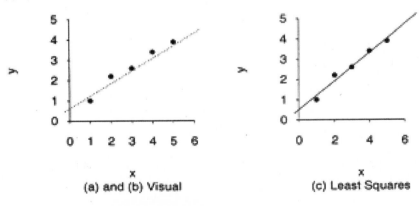

(a) and (b) Visual   (c) Least Squares

(c) We use the alternative form of calculation

x	y	xy	x^2
1	1.0	1.0	1
2	2.2	4.4	4
3	2.6	7.8	9
4	3.4	13.6	16
5	3.9	19.5	25
15	13.1	46.3	55
$\bar{x}=3$	$\bar{y}=2.62$	$\sum xy$	$\sum x^2$

so

$$S_{xy} = \sum xy - \frac{(\sum x)(\sum y)}{n} = 46.3 - \frac{(15)(13.1)}{5} = 7.00$$

$$S_{xx} = \sum x^2 - \frac{(\sum x)^2}{n} = 55 - \frac{(15)^2}{5} = 10.00$$

and

$$\hat{\beta}_1 = \frac{S_{xy}}{S_{xx}} = \frac{7}{10} = 0.70$$

$$\hat{\beta}_0 = \bar{y} - \hat{\beta}_1\bar{x} = 2.62 - (0.70)3 = 0.52$$

and the least squares line is $\hat{y} = 0.52 + 0.70x$. This is the solid line in the figure.

3.41 (a)

$$\hat{\beta}_1 = \frac{S_{xy}}{S_{xx}} = \frac{10.2}{9.4} = 1.085$$

$$\hat{\beta}_0 = \bar{y} - \hat{\beta}_1\bar{x} = \frac{39.9}{9} - (1.085)\frac{19}{9} = 2.143$$

(b) $\hat{y} = 2.143 + 1.085(3) = 5.398$ or about 5.4 minutes.

3.43 (a) The scatter diagram is shown below.

Amount of Garbage for Ten-Year Periods

Y = -176.888 + 1.47712X

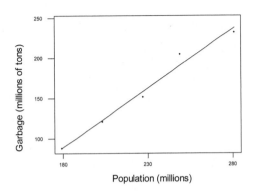

(b) We calculate $n = 5$, $\sum x = 1139$, $\sum y = 798$, $S_{xx} = 6276.8$ and $S_{xy} = 9271.6$.

$$\hat{\beta}_1 = \frac{S_{xy}}{S_{xx}} = \frac{9271.6}{6276.8} = 1.477$$

$$\hat{\beta}_0 = \bar{y} - \hat{\beta}_1 \bar{x} = \frac{798}{5} - (1.477)\frac{1139}{5} = -176.89$$

so the least squares line is $\hat{y} = -176.89 + 1.48x$.

(c) The slope of the least squares line, 1.48, says a 1.48 million tons of garbage is created for each 1 million people. Hence, each person generates approximately 1.48 tons of garbage.

3.45 (a) $\bar{x} = 542/30 = 18.067$

(b) The relative frequencies, by manufacturer, are:

Manufacturer	Carbohydrates		Total
	Above mean	Below mean	
General Mills	4	6	10
Kellogg	4	6	10
Quaker	4	6	10
Total	12	18	30

(c)

Manufacturer	Carbohydrates		Total
	Above mean	Below mean	
General Mills	0.4	0.6	1.0
Kellogg	0.4	0.6	1.0
Quaker	0.4	0.6	1.0

The row proportions are exactly the same for each row.

3.47 (a) The frequency table is

Size	Drive		Total
	2-Wheel	4-Wheel	
Small	12	21	33
Full	27	15	42
Total	39	36	75

(b) The relative frequencies are:

Size	Drive		Total
	2-Wheel	4-Wheel	
Small	0.16	0.28	0.44
Full	0.36	0.20	0.56
Total	0.52	0.48	1.00

(c)

Size	Drive		Total
	2-Wheel	4-Wheel	
Small	0.364	0.636	1.000
Full	0.643	0.357	1.000

(d) A larger proportion of small truck purchasers prefer 4-wheel drive.

3.49 (a) Negative; typically, the more time spent on the computer the fewer hours available for friends and other activities.

(b) Somewhat negative; most students cram for finals and the more exams the more late night studying during finals and the fewer hours of sleep.

(c) No relation.

(d) Positive; higher temperature tends to make people more thirsty.

3.51 (a) The scatter diagram is shown below.

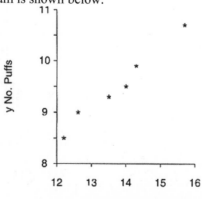

(b) With $n = 6$, we calculate

$$\sum x = 82.3 \quad , \quad \sum y = 56.9$$

$$\sum x^2 = 1,136.83 \quad \sum y^2 = 785.21 \quad \sum xy = 542.49$$

so

$$S_{xx} = \sum x^2 - \frac{(\sum x)^2}{n} = 1,136.83 - \frac{(82.3)^2}{6} = 7.948$$

$$S_{yy} = \sum y^2 - \frac{(\sum y)^2}{n} = 542.49 - \frac{(56.9)^2}{6} = 2.888$$

$$S_{xy} = \sum xy - \frac{(\sum x)(\sum y)}{n} = 785.21 - \frac{(82.3)(56.9)}{6} = 4.732$$

Consequently,

$$r = \frac{S_{xy}}{\sqrt{S_{xx}}\sqrt{S_{yy}}} = \frac{4.732}{\sqrt{7.948}\sqrt{2.888}} = 0.988.$$

3.53 (a)

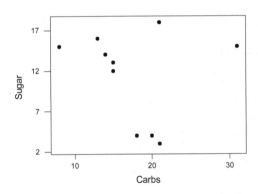

(b) With $n = 8$, we calculate

$$\sum x = 114 \quad , \quad \sum y = 176$$

$$\sum x^2 = 1580 \quad \sum y^2 = 3446 \quad \sum xy = 1957$$

so

$$S_{xx} = \sum x^2 - \frac{(\sum x)^2}{n} = 1580 - \frac{(114)^2}{10} = 280.4$$

$$S_{yy} = \sum y^2 - \frac{(\sum y)^2}{n} = 3446 - \frac{(176)^2}{10} = 348.4$$

$$S_{xy} = \sum xy - \frac{(\sum x)(\sum y)}{n} = 1957 - \frac{(114)(176)}{10} = -49.4$$

Consequently,

$$r = \frac{S_{xy}}{\sqrt{S_{xx}}\sqrt{S_{yy}}} = \frac{-49.4}{\sqrt{280.4}\sqrt{348.4}} = -0.158.$$

The scatter plot and the small value of r indicate sugar content and carbohydrate content are unrelated in Kellogg's cereals.

3.55 (a) (b) The scatter plot and visually drawn line are shown below.

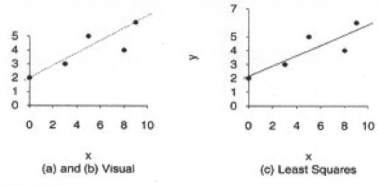

(a) and (b) Visual   (c) Least Squares

(c) We calculate

x'	y'	$x'-\overline{x'}$	$y'-\overline{y'}$	$(x'-\overline{x'})(y'-\overline{y'})$	$(x'-\overline{x'})^2$
0	2	−5	−2	10	25
3	3	−2	−1	2	4
5	5	0	1	0	0
8	4	3	0	0	9
9	6	4	2	8	16
25	20	0	0	20	54
$\overline{x'}=5$	$\overline{y'}=5$			$S_{x'y'}$	$S_{x'x'}$

so

$$\hat{\beta}_1 = \frac{S_{xy}}{S_{xx}} = \frac{20}{54} = 0.370$$

$$\hat{\beta}_0 = \overline{y} - \hat{\beta}_1\overline{x} = 4 - (0.370)5 = 2.15$$

and the least squares line is $\hat{y} = 2.15 + 0.370x$. This is the solid line in the figure.

3.57 (a) x = road roughness and y = gas consumption.
   (b) x = number of wins and y = total sales.
   (c) x = trip distance and y = number of weekends at home.

3.59 (a)

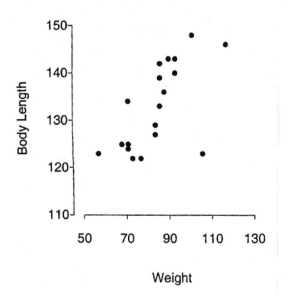

(b)

```
Correlations: bodyleng, weight


Pearson correlation of bodyleng and weight = 0.649
```

3.61 (a) The scatter diagram is shown below.

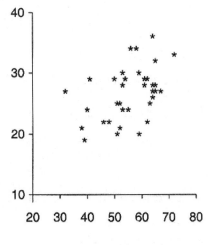

(b) With $n = 36$, we calculate

$$\sum x = 2006 \quad , \quad \sum y = 961$$

$$\sum x^2 = 114,950 \quad \sum xy = 54,166 \quad \sum y^2 = 26,281$$

so

$$S_{xx} = \sum x^2 - \frac{\left(\sum x\right)^2}{n} = 114,950 - \frac{(2006)^2}{36} = 3,171.22$$

$$S_{yy} = \sum y^2 - \frac{\left(\sum y\right)^2}{n} = 26,281 - \frac{(961)^2}{36} = 627.64$$

$$S_{xy} = \sum xy - \frac{\left(\sum x\right)\left(\sum y\right)}{n} = 54,166 - \frac{(2006)(961)}{36} = 616.94$$

Consequently,

$$r = \frac{S_{xy}}{\sqrt{S_{xx}}\sqrt{S_{yy}}} = \frac{616.94}{\sqrt{3,171.22}\sqrt{627.64}} = 0.437.$$

Chapter 4

PROBABILITY

4.1 (a) (ii), (v)   (e) (v)
    (b) (iv), (v)   (f) (iii), (v)
    (c) (vi)   (g) (i)
    (d) (vi)

4.3 (a) (i)   (b)   (iii)   (c)   (ii)

4.5 (a) {0, 1}

    (b) {0, 1,..., 344}

(c) $\{t : 90 < t < 425.4\}$

4.7 (a) Let us identify Bob, John, Linda, and Sue by their initials B, J, L, and S,
    respectively. We make a tree diagram:

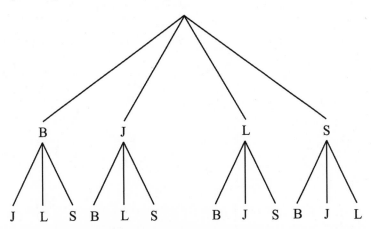

S = {BJ, BL, BS, JB, JL, JS, LB, LJ, LS, SB, SJ, SL}
(b) A = {LB, LJ, LS}, B = {JL, LJ, JS, SJ, LS, SL}

4.9 $P(e_1) + P(e_2) + P(e_3) = 0.2 + 0.5 + 0.1 = 0.8$. Since $P(S) = 1$, we must have
$P(e_4) = 1 - 0.8 = 0.2$.

4.11 (a) yes
      (b) no, because the sum of probabilities is greater than 1
      (c) yes

4.13 Denote May by e_1 and so on. Because $1 + 3 + 6 + 10 = 20$, we have

$$P(e_1) = \frac{1}{20}, \ P(e_2) = \frac{3}{20}, \ P(e_3) = \frac{6}{20}, \ P(e_4) = \frac{10}{20}$$

so that $P(A) = P(e_1) + P(e_2) = \dfrac{4}{20} = 0.2$.

4.15 The relative frequencies are based on a very large number of cases and will
      therefore be very good approximations to the probabilities. Since
$$P(\text{weekday}) + P(\text{weekend}) = 1$$
      and $P(\text{weekend})$ is approximately 0.257, the probability of a weekday birth is
      approximately $1 - 0.257 = 0.743$.

4.17 (a) The tree diagram is

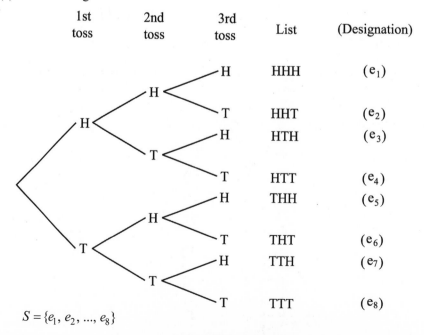

1st toss	2nd toss	3rd toss	List	(Designation)
		H	HHH	(e_1)
	H	T	HHT	(e_2)
H		H	HTH	(e_3)
	T	T	HTT	(e_4)
		H	THH	(e_5)
	H	T	THT	(e_6)
T		H	TTH	(e_7)
	T	T	TTT	(e_8)

$S = \{e_1, e_2, ..., e_8\}$

(b) Assuming the coins are all fair, all the elementary outcomes are equally likely.

$$P(e_1) = P(e_2) = \cdots = P(e_8) = \frac{1}{8}$$

(c) [Exactly one head] $= \{e_4, e_6, e_7\}$, its probability is $\frac{3}{8}$.

4.19 (a) Let e_1, e_2, and e_3 denote the outcomes of getting a ticket numbered 1, 2, and 3, respectively. Then $S = \{e_1, e_2, e_3\}$. Since all 8 tickets are equally likely to be drawn, and there are 2 tickets with number 1, we have $P(e_1) = \frac{2}{8}$. Likewise,

$P(e_2) = \frac{3}{8}$ and $P(e_3) = \frac{3}{8}$.

(b) [Odd-numbered ticket drawn] $= \{e_1, e_3\}$, so the probability is

$$P(e_1) + P(e_3) = \frac{2}{8} + \frac{3}{8} = \frac{5}{8} = 0.625 .$$

4.21 (a) Each elementary outcome is a pair of numbers, the first corresponds to the white die and the second to the colored die.

$$A = \{(1, 5), (2, 4), (3, 3), (4, 2), (5, 1)\}$$
$$B = \{(1, 6), (2, 5), (3, 4), (4, 3), (5, 2), (6, 1)\}$$
$$C = \{(2, 6), (4, 6), (6, 6), (1, 5), (3, 5), (5, 5),$$
$$(2, 4), (4, 4), (6, 4), (1, 3), (3, 3), (5, 3),$$
$$(2, 2), (4, 2), (6, 2), (1, 1), (3, 1), (5, 1)\}$$
$$D = \{(1, 1), (2, 2), (3, 3), (4, 4), (5, 5), (6, 6)\} .$$

(b) Probability $\frac{1}{36}$ for each elementary outcome.

(c) $P(A) = \frac{5}{36}$, $P(B) = \frac{6}{36} = \frac{1}{6}$, $P(C) = \frac{18}{36} = \frac{1}{2}$, $P(D) = \frac{6}{36} = \frac{1}{6}$

4.23 (a) The tree diagram is

Answer to Q_1	Answer to Q_2	List
T	T	TT
	F	TF
	I	TI
F	T	FT
	F	FF
	I	FI
I	T	IT
	F	IF
	I	II

So, S={TT, TF, TI, FT, FF, FI, IT, IF, II}.

(b) Because the student selects the answers at random, the 9 elementary outcomes in S are all equally likely, each has a probability $\frac{1}{9}$. Let us suppose that the correct answers are T for Q_1 and T for Q_2. Then, the event "one correct answer" has the composition {TF, TI, FT, IT}, so that

$$P(\text{one correct answer}) = \frac{4}{9}.$$

Note: Whatever be the correct answers for Q_1 and Q_2, there will be four cases in which one marked answer will match and one will not match.

4.25 (a) The 15 persons are equally likely to be selected. Among them there is only one of blood group AB, so that $P[AB] = \frac{1}{15}$.

(b) The number of persons of blood group either A or B is $5+6=11$, so that the required probability is $\frac{11}{15}$.

(c) $P[\text{not O}] = \frac{5+6+1}{15} = \frac{12}{15}$.

4.27 $S = \{N, YN, YYN, YYYN, YYYYN, YYYYY\}$

4.29 (a) Letting c, b, and v denote "compliance", "borderline case", and "violation", respectively,
$$S = \{c_1, c_2, ..., c_9, b_1, b_2, b_3, v_1, v_2\}.$$

(b) The 14 elementary outcomes are equally likely, and two of them, namely v_1 and v_2, constitute the event that a violator is detected. The probability is $2/14 = 0.143$.

4.31 (a) The successive days cannot be considered as independent trials. The rate on one day is the same as, or is very close to, the rate on the next day. The results for successive days are not independent. There may also be a trend in rates over the year.

(b) Cars brought in with other problems are more likely to have an emission problem. For instance, there would be too many old cars in this sample.

(c) Not a representative sample. Since most air conditioners are sold during the summer, the observed relative frequency would be too small.

4.33 (a) Since the gift certificates are assigned at random, all 6 elementary outcomes are equally likely, so that each has probability 1/6.

   (b) $P(A) = \dfrac{3}{6} = \dfrac{1}{2}, \qquad P(B) = \dfrac{2}{6} = \dfrac{1}{3}.$

4.35 (a) The Venn diagram is

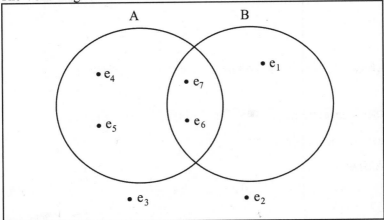

Figure 4.1: Venn Diagram for Exercise 4.35(a)

   (b) (i) $AB = \{e_6, e_7\}$

      (ii) $\bar{B} = \{e_2, e_3, e_4, e_5\}$

      (iii) $A\bar{B} = \{e_4, e_5\}$

      (iv) $A \cup B = \{e_1, e_4, e_5, e_6, e_7\}$

4.37 (a) $\bar{C} = \{e_1, e_2, e_3, e_4, e_5, e_7\}, \quad P(\bar{C}) = 0.7$

   (b) $AB = \{e_2, e_6, e_7\}, \qquad P(AB) = 0.12 + 0.15 + 0.15 = 0.42$

   (c) $A\bar{B} = \{e_1, e_5\}, \qquad P(A\bar{B}) = 0.04 + 0.15 = 0.19$

   (d) $\bar{A}\bar{C} = \{e_3, e_4\}, \qquad P(\bar{A}\bar{C}) = 0.12 + 0.12 = 0.24$

4.39 (a) Denote by $e_1, e_2, e_3,$ and e_4 the elementary outcomes that the person hired is candidate number 1, 2, 3, and 4, respectively. The Venn diagram is given in Figure 4.2.

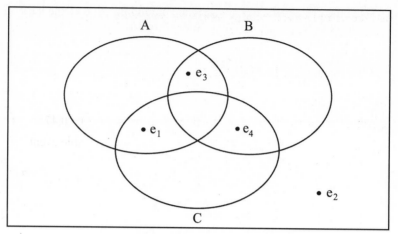

Figure 4.2: Venn diagram for Exercise 4.39(a)

(b) $A \cup B = \{e_1, e_3, e_4\}$ and $AB = \{e_3\}$

4.41 (a) $P(A) = 0.06 + 0.1 + 0.09 = 0.25$ and $P(B) = 0.06 + 0.1 + 0.09 + 0.09 = 0.34$
   Also, since $AB = \{e_5, e_8\}$, $P(AB) = 0.1 + 0.09 = 0.19$.

   (b) $P(A \cup B) = P(A) + P(B) - P(AB) = 0.25 + 0.34 - 0.19 = 0.40$

   (c) Since $A \cup B = \{e_1, e_5, e_8, e_2, e_9\}$,
   $$P(A \cup B) = 0.06 + 0.1 + 0.09 + 0.06 + 0.09 = 0.40$$

   (d) $P(\overline{B}) = 1 - P(B) = 1 - 0.34 = 0.66$. Alternatively, since $\overline{B} = \{e_1, e_3, e_4, e_6, e_7\}$,
   $$P(\overline{B}) = 0.06 + 0.1 + 0.1 + 0.2 + 0.2 = 0.66.$$

4.43 (a) The specified probabilities are entered in the table and underlined. The other
   entries are obtained in part (b).

	B	$\overline{B}$	
A	0.20	0.32	0.52
$\overline{A}$	0.16	0.32	0.48
	0.36	0.64	1.00

   (b) Since $A = AB \cup A\overline{B}$, a union of mutually exclusive events, we have
   $P(A) = P(AB) + P(A\overline{B})$. Hence, for the given information,
   $0.52 = 0.20 + P(A\overline{B})$. Thus, solving for $P(A\overline{B})$ yields:
   $$P(A\overline{B}) = 0.52 - 0.20 = 0.32.$$
   Similarly,
   $$P(\overline{A}B) = 0.36 - 0.20 = 0.16, \quad P(\overline{A}\overline{B}) = 1 - (0.20 + 0.32 + 0.16) = 0.32$$

4.45 (a) The completed probability table is given below

	B	$\bar{B}$	
A	0.20	0.12	0.32
$\bar{A}$	0.15	0.53	0.68
	0.35	0.65	1.00

(b) $P(A\bar{B}) = 0.12$

(c) $P(A \cup B) = P(A) + P(B) - P(AB) = 0.32 + 0.35 - 0.20 = 0.47$

(d) $P(A\bar{B} \cup \bar{A}B) = P(A\bar{B}) + P(\bar{A}B)$ (union of incompatible events)
$$= 0.12 + 0.15 = 0.27$$

4.47 (a) $P(\bar{A}) = 0.4 + 0.25 = 0.65$

(b) $P(A\bar{B}) = 0.15$

(c) $P(A\bar{B}) + P(\bar{A}B) = 0.15 + 0.4 = 0.55$

4.49 Denoting 'violation' by V and 'compliance' by C, the classification of the 18 restaurants is shown in the following table:

		Safety		
		V	C	Total
Sanitary	V	4	3	7
	C	4	7	11
		8	10	18

$$P(CC) = \frac{7}{18} = 0.389$$

4.51 (a) $P(A) = 0.08 + 0.02 + 0.20 + 0.10 = 0.40$
$P(B) = 0.15 + 0.10 + 0.08 + 0.02 = 0.35$
$P(BC) = 0.15 + 0.08 = 0.23$
$P(ABC) = 0.08$

(b) (i) Light case and above 40.

$$P(\bar{A}\bar{B}) = 0.15 + 0.20 = 0.35$$

(ii) Either a light case or the parents are not diabetic or both.

$$P(\bar{A} \cup \bar{C}) = 0.15 + 0.10 + 0.15 + 0.20 + 0.02 + 0.10 = 0.72$$

(iii) A light case, age is below 40 and parents are not diabetic.

$$P(\bar{A}B\bar{C}) = 0.10.$$

4.53 (a) With the stated numbers identifying the gift boxes, the list is:

$$(1,1), \quad (1,2), \quad (1,3), \quad (1,4), \quad (1,5)$$

$$(2,1), \quad (2,2), \quad (2,3), \quad (2,4), \quad (2,5)$$

$$(3,1), \quad (3,2), \quad (3,3), \quad (3,4), \quad (3,5)$$

$$(4,1), \quad (4,2), \quad (4,3), \quad (4,4), \quad (4,5)$$

$$(5,1), \quad (5,2), \quad (5,3), \quad (5,4), \quad (5,5)$$

The 25 elementary outcomes are equally likely, so each has the probability $\frac{1}{25}$.

(b) $A = \{(2,1), \ (2,2), \ (2,3), \ (2,4), \ (2,5), (3,1), \ (3,2), \ (3,3), \ (3,4), \ (3,5),$
$\quad (1,2), \ (4,2), \ (5,2), \ (1,3), \ (4,3), \ (5,3)\},$
So, $P(A) = \frac{16}{25}$.

$B = \{(3,1), \ (3,2), \ (3,3), \ (3,4), \ (3,5), (4,1), \ (4,2), \ (4,3), \ (4,4), \ (4,5),$
$\quad (5,1), \ (5,2), \ (5,3), \ (5,4), \ (5,5), (1,3), \ (1,4), \ (1,5), \ (2,3), \ (2,4), \ (2,5)\}$
So, $P(B) = \frac{21}{25}$.

$AB = \{(2,3), \ (2,4), \ (2,5), \ (1,3), (3,1), \ (3,2), \ (3,3), \ (3,4), \ (3,5),$
$\quad (4,2), \ (4,3), \ (5,2), \ (5,3)\},$
So, $P(AB) = \frac{13}{25}$.

4.55 It is reasonable to assume that a person with a college degree is more likely to command a higher salary than a person without a college degree, so we would expect $P(A \mid B) > P(A)$. If A and B are independent, then $P(A \mid B) = P(A)$ so by the previous reasoning, A and B are not independent.

4.57 (a) $P(B \mid A) = \dfrac{P(AB)}{P(A)} = \dfrac{0.35}{0.68} = 0.515$

(b) From $P(A) = 0.68$ and $P(AB) = 0.35$ and the fact that $A = AB \cup A\bar{B}$ (union of mutually exclusive events), we calculate $P(A\bar{B}) = 0.68 - 0.35 = 0.33$.
Therefore, since $P(\bar{B} \mid A) = \dfrac{P(A\bar{B})}{P(A)}$, we have $P(\bar{B} \mid A) = \dfrac{0.33}{0.68} = 0.485$.

Alternatively, note that B and $\overline{B}$ are complementary events so that

$$P(\overline{B} \mid A) = 1 - P(B \mid A) = 1 - 0.515 = 0.485$$

(c) Observe that $P(\overline{A}) = 1 - P(A) = 1 - 0.68 = 0.32$ and

$$P(\overline{A}B) = P(B) - P(AB) = .55 - .35 = .20 \text{ (since } B = \overline{A}B \cup AB \text{ as in (b)).}$$

Therefore, since $P(B \mid \overline{A}) = \dfrac{P(\overline{A}B)}{P(\overline{A})}$, we have $P(B \mid \overline{A}) = \dfrac{.20}{.32} = .625$.

4.59 Observe that $P(B \mid A) = \dfrac{P(AB)}{P(A)} = \dfrac{0.001}{0.101} = 0.0099$ and $P(B) = 0.05 + 0.001 = 0.051$.

Since $P(B) \neq P(B \mid A)$, the events A and B are not independent.

4.61 (a) $P(\overline{A}) = 1 - P(A) = 1 - 0.4 = 0.6$

(b) $P(AB) = P(B)P(A \mid B) = 0.25 \times 0.7 = 0.175$

(c) $P(A \cup B) = P(A) + P(B) - P(AB) = 0.4 + 0.25 - 0.175 = 0.475$

4.63

2	Green
3	Red

$\rightarrow 2$, without replacement

(a) We denote G for green, R for red and attach subscripts to identify the order of the draws. Since the event A, a green ball appears in the first draw, has nothing to do with the second draw, we identify $A = G_1$ so

$$P(A) = P(G_1) = \frac{2}{5} = 0.4$$

The event $B = G_2$ is the union of $G_1 G_2$ and $R_1 G_2$.

$$P(G_1 G_2) = P(G_1)P(G_2 \mid G_1) = \frac{2}{5} \times \frac{1}{4} = \frac{2}{20}$$

$$P(R_1 G_2) = P(R_1)P(G_2 \mid R_1) = \frac{3}{5} \times \frac{2}{4} = \frac{6}{20}$$

Hence, $P(B) = P(G_2) = \dfrac{2}{20} + \dfrac{6}{20} = \dfrac{8}{20} = 0.4$

(b) $P(AB) = P(G_1G_2) = \dfrac{2}{20} = 0.1$. On the other hand, $P(A)P(B) = 0.4 \times 0.4 = 0.16$,

and this is different from $P(AB)$. Therefore, A and B are not independent.

4.65 We use the symbols M for male, F for female, U for unemployed and E for employed.

(a) $P(M) = 0.6$
$P(U \mid M) = 0.051$
$P(U \mid F) = 0.043$

(b) $P(UM) = 0.051 \times 0.6 = 0.0306$ and $P(UF) = 0.043 \times (1 - 0.6) = 0.0172$
Adding these we obtain $P(U) = 0.0306 + 0.0172 = 0.0478$, so the overall rate of unemployment is 4.8%.

(c) To find $P(F \mid U)$, we use the results $P(U) = 0.0478$ and $P(UF) = 0.0172$ and obtain $P(F \mid U) = \dfrac{P(FU)}{P(U)} = \dfrac{0.0172}{0.0478} = 0.360$.

4.67 (a) If independent, $P(AB) = P(A)P(B) = 0.6 \times 0.22 = 0.132$ so
$P(A \cup B) = P(A) + P(B) - P(AB) = 0.6 + 0.22 - 0.132 = 0.688$

(b) If mutually exclusive, $P(AB) = 0$ so

$$P(A \cup B) = P(A) + P(B) = 0.6 + 0.22 = 0.82$$

(c) If A and B are mutually exclusive, we have $A\bar{B} = A$ so

$$P(A \mid \bar{B}) = \dfrac{P(A\bar{B})}{P(\bar{B})} = \dfrac{0.6}{0.78} = 0.769.$$

4.69 We use the symbol D for defective and G for good, and attach subscripts to identify the order of selection.

(a) Here the second selection is irrelevant. If one air conditioner is selected at random, the probability of its being defective is $P(D_1) = \dfrac{3}{15} = 0.2$.

(b) $P(D_1G_2) = P(D_1)P(G_2 \mid D_1) = \dfrac{3}{15} \times \dfrac{12}{14} = \dfrac{36}{210} = 0.171$.

(c) $P(D_1D_2) = P(D_1)P(D_2 \mid D_1) = \dfrac{3}{15} \times \dfrac{2}{14} = \dfrac{6}{210} = 0.0286$

(d) The event D_2 is the union of two incompatible events:

$D_2 = D_1D_2 \cup G_1D_2$ so $P(D_2) = P(D_1D_2) + P(G_1D_2)$.

We have already calculated $P(D_1D_2)$ in part (c). In the same way

$P(G_1D_2) = P(G_1)P(D_2 \mid G_1) = \dfrac{12}{15} \times \dfrac{3}{14} = \dfrac{36}{210}$. Therefore,

$$P(D_2) = \dfrac{6}{210} + \dfrac{36}{210} = \dfrac{42}{210} = \dfrac{3}{15} = 0.2.$$

(e) [exactly one defective] $= D_1G_2 \cup G_1D_2$.

We have already found that $P(D_1G_2) = P(G_1D_2) = \dfrac{36}{210}$.

Hence, $P(\text{exactly one defective}) = \dfrac{36}{210} + \dfrac{36}{210} = \dfrac{72}{210} = 0.343$.

4.71 The classification of the 20 rats is shown in the following table:

	Infected (I)	Not infected (N)	Total
Male (M)	7	5	12
Female (F)	2	6	8
	9	11	20

(a) There are 9 infected rate of which 2 are females. Therefore, $P(F \mid I) = \dfrac{2}{9}$.

Alternatively, we can use the definition of conditional probability

$$P(F \mid I) = \dfrac{P(FI)}{P(I)} = \dfrac{2/20}{9/20} = \dfrac{2}{9}.$$

(b) There are 12 males of which 7 are infected, so $P(I \mid M) = \dfrac{7}{12}$.

(c) $P(I) = \dfrac{9}{20}$, $P(M) = \dfrac{12}{20}$, $P(IM) = \dfrac{7}{20} = 0.35$

$P(I)P(M) = \dfrac{9}{20} \times \dfrac{12}{20} = 0.27$

Since $P(IM) \neq P(I)P(M)$, the events are not independent.

4.73 (a) BC, $P(BC) = 0$ since B and C are incompatible.

(b) $A \cup B$, $P(A \cup B) = P(A) + P(B) - P(AB)$

Now, by independence, $P(AB) = P(A)P(B) = 0.7 \times 0.2 = 0.14$

Hence, $P(A \cup B) = 0.7 + 0.2 - 0.14 = 0.76$

(c) $\bar{B}$, $P(\bar{B}) = 1 - P(B) = 1 - 0.2 = 0.8$

(d) ABC, $P(ABC) = 0$ since B and C are incompatible and so are AB and C.

4.75 Denote the events by:

S = the cooling system functions
S_1 = the primary unit functions
F_1 = the primary unit fails
$S_2(F_2)$ = the back-up unit functions (fails)

Then S can be expressed as the union of two incompatible events: $S = S_1 \cup (F_1 S_2)$
As such, $P(S) = P(S_1) + P(F_1 S_2)$. Next, observe that $P(S_1) = .999$ and

$$P(F_1 S_2) = P(F_1)P(S_2 \mid F_1) = 0.001 \times 0.910 = 0.00091.$$

Therefore, $P(S) = 0.999 + 0.00091 = 0.99991$.

4.77 (a) Denoting the success and failure in each test by S and F, respectively, the
         sample space is conveniently listed with the tree diagram below.

Test 1	Test 2	Test 3	List	Probability
		S	SSS	0.512
	S	F	SSF	0.128
S	F	S	SFS	0.128
		F	SFF	0.032
	S	S	FSS	0.128
F		F	FSF	0.032
	F	S	FFS	0.032
		F	FFF	0.008
				1.000

To calculate the probabilities we note that $P(S) = 0.8$ and $P(F) = 1 - 0.8 = 0.2$.
Since the tests are independent, we calculate

$$P(SSS) = P(S)P(S)P(S) = 0.8 \times 0.8 \times 0.8 = 0.512$$
$$P(SSF) = P(S)P(S)P(F) = 0.8 \times 0.8 \times 0.2 = 0.128,\ \text{etc.}$$

The results are shown in the column of probability in the above figure.

(b) P(at least two successes) = P(2 successes) + P(3 successes)
                    $= (0.128 + 0.128 + 0.128) + 0.512 = 0.896$.

4.79 (a) The following probabilities are specified:

$$P(D) = 0.03, \qquad P(N) = 1 - P(D) = 1 - 0.03 = 0.97$$
$$P(+\,|\,N) = 0.10, \quad P(-\,|\,N) = 1 - P(+\,|\,N) = 1 - 0.10 = 0.90$$
$$P(-\,|\,D) = 0.05, \quad P(+\,|\,D) = 1 - P(-\,|\,D) = 1 - 0.05 = 0.95$$

Using the above probabilities, we calculate

$$P(D+) = P(D)P(+\,|\,D) = 0.03 \times 0.95 = 0.0285$$
$$P(D-) = P(D)P(-\,|\,D) = 0.03 \times 0.05 = 0.0015$$
$$P(N+) = P(N)P(+\,|\,N) = 0.97 \times 0.10 = 0.0970$$
$$P(N-) = P(N)P(-\,|\,N) = 0.97 \times 0.90 = 0.8730$$

These are entered in the following probability table:

	+	−	Total
D	0.0285	0.0015	0.0300
N	0.0970	0.8730	0.9700
	0.1255	0.8745	1.0000

(b) The required probability is the conditional probability of D given $+$

$$P(D\,|\,+) = \frac{P(D+)}{P(+)} = \frac{0.0285}{0.1255} = 0.227.$$

4.81 Denote the events

$$S = \text{strep-throat} \quad \text{and} \quad A = \text{Allergy}$$

We are given that $P(S) = 0.25$, $P(A) = 0.4$, and $P(SA) = 0.1$.

(a) $P(S \cup A) = P(S) + P(A) - P(SA) = 0.25 + 0.4 - 0.1 = 0.55$

(b) $P(S)P(A) = 0.25 \times 0.4 = 0.10 = P(SA)$, so the events are independent.

4.83 (a) $\dbinom{6}{3} = \dfrac{6 \times 5 \times 4}{3 \times 2 \times 1} = 20$

(b) $\dbinom{10}{3} = \dfrac{10 \times 9 \times 8}{3 \times 2 \times 1} = 120$

(c) $\dbinom{22}{2} = \dfrac{22 \times 21}{2 \times 1} = 231$

(d) $\dbinom{22}{20} = \dbinom{22}{2} = 231$ (see part c)

(e) $\dbinom{30}{3} = \dfrac{30 \times 29 \times 28}{3 \times 2 \times 1} = 4060$

(f) $\dbinom{30}{27} = \dbinom{30}{3} = 4060$ (see part e)

4.85 (a) The number of possible selections of 4 persons out of 10 persons is
$$\binom{10}{4} = \frac{10 \times 9 \times 8 \times 7}{4 \times 3 \times 2 \times 1} = 210$$

(b) The number of possible selections of 2 men out of 6 men is
$$\binom{6}{2} = \frac{6 \times 5}{2 \times 1} = 15$$

and the number of possible selections of 2 women out of 4 women is
$$\binom{4}{2} = \frac{4 \times 3}{2 \times 1} = 6$$

Since the men can be selected in 15 ways, and for each selection of men, there are 6 ways the women can be selected, the number of possible selections of two men and two women is $15 \times 6 = 90$.

4.87 (a) The number of possible selections of 5 children out of 11 is $\dbinom{11}{5} = 462$.

(b) The number of selections of 2 out of the 4 young males is $\dbinom{4}{2} = 6$.

The number of selections of 3 out of the 7 young females is $\dbinom{7}{3} = 35$.

Each of the 6 choices of the young males can accompany each of the 35 choices of the young females, so the number of selections of 2 males and 3 female students is
$$\binom{4}{2} \times \binom{7}{3} = 6 \times 35 = 210.$$

4.89 The number of possible samples of 5 jurors out of 17 is $\binom{17}{5} = 6188$. Under random sampling, these 6188 possible samples are equally likely. The number of possible samples where all 5 jurors selected are males is $\binom{10}{5} = 252$. The jury selection shows discrimination since

$$P(\text{no female members}) = \frac{\binom{10}{5}}{\binom{17}{5}} = \frac{252}{6188} = 0.041.$$

4.91 The batch contains 4 defective and 16 good alternators. The number of possible samples of size 3 is $\binom{20}{3} = 1140$ and these are equally likely.

(a) $P(A) = \dfrac{\binom{16}{3}}{\binom{20}{3}} = \dfrac{560}{1140} = 0.491$

(b) $P(B) = \dfrac{\binom{4}{2} \times \binom{16}{1}}{\binom{20}{3}} = \dfrac{6 \times 16}{1140} = 0.084$.

4.93 No. The states with fewer seniors get more representation in this process than what the random selection would permit.

4.95 (a) The number of possible selections of 3 plots out of 9 is $\binom{9}{3} = 84$, and the 84 selections are equally likely. One row can be chosen in $\binom{3}{1} = 3$ ways, and within that row, 3 plots can be chosen in $\binom{3}{3} = 1$ way. Therefore, the number of choices such that the three plots are in the same row is $\binom{3}{1} \times \binom{3}{3} = 3 \times 1 = 3$.

Hence, the required probability $= \dfrac{3}{84} = 0.036$.

(b) The number of ways one plot can be selected from row 1 is $\binom{3}{1} = 3$. Likewise, a plot can be selected from row 2 in 3 ways, and from row 3 in 3 ways. The number of possible selections of 3 plots, one in each row, is $3 \times 3 \times 3 = 27$, so the required probability $= \dfrac{27}{84} = 0.321$.

4.97

Row A			Row B		
4	bushy		6	bushy	
4	lean	$\rightarrow 2$	3	lean	$\rightarrow 2$
8			9		

(a) There are 8 trees in row A from which 2 trees can be selected in $\binom{8}{2} = 28$ ways.

Of these 28 equally likely selections, there are $\binom{4}{2} = 6$ selections in which both trees are bushy. Therefore,

$$P[2 \text{ bushy trees selected in row A}] = \frac{6}{28}.$$

Similarly,

$$P[2 \text{ lean trees selected in row B}] = \frac{\binom{3}{2}}{\binom{9}{2}} = \frac{3}{36}$$

By independence of the two selections, the required probability is therefore

$$= \frac{6}{28} \times \frac{3}{36} = 0.018.$$

(b) Let A_0, A_1, and A_2 respectively denote the events of getting exactly 0, 1, or 2 bushy trees in row A, and let B_0, B_1, B_2 denote the corresponding events for row B. Then

[Exactly 2 bushy] $= A_2 B_0 \cup A_1 B_1 \cup A_0 B_2$ (union of mutually exclusive events)

$$P(A_2 B_0) = P(A_2)P(B_0) = \frac{\binom{4}{2}}{\binom{8}{2}} \times \frac{\binom{3}{2}}{\binom{9}{2}} = \frac{6}{28} \times \frac{3}{36} = 0.018 \quad (\text{see part (a)})$$

$$P(A_1B_1) = P(A_1)P(B_1) = \frac{\binom{4}{1}\binom{4}{1}}{\binom{8}{2}} \times \frac{\binom{6}{1}\binom{3}{1}}{\binom{9}{2}} = \frac{16}{28} \times \frac{18}{36} = 0.286$$

$$P(A_0B_2) = P(A_0)P(B_2) = \frac{\binom{4}{2}}{\binom{8}{2}} \times \frac{\binom{6}{2}}{\binom{9}{2}} = \frac{6}{28} \times \frac{15}{36} = 0.089$$

Adding these probabilities we get $P[\text{exactly 2 bushy}] = 0.393$.

4.99

5	below thirty
7	over thirty

$\quad$ 12 $\qquad$ $\rightarrow 4$ randomly selected

(a) The number of possible choices of 4 persons out of 12 is $\binom{12}{4} = 495$.

(b) The number of choices of 3 persons below thirty and 1 over thirty is

$$\binom{5}{3} \times \binom{7}{1} = 10 \times 7 = 70.$$

So, the required probability $= \frac{70}{495} = 0.141$.

4.101

6	yellow
5	red

$\quad$ 11 $\qquad$ $\rightarrow 4$

The number of possible samples of 4 bulbs out of 11 is $\binom{11}{4} = 330$, and all choices are equally likely.

(a) The number of ways 2 red and 2 yellow bulbs can be selected is

$$\binom{5}{2} \times \binom{6}{2} = 10 \times 15 = 150$$

So, $P[\text{exactly 2 red}] = \frac{150}{330} = 0.455$.

(b) We calculate

$$P[2 \text{ red}] = \frac{150}{330} \text{ (done in part (a))}$$

$$P[3 \text{ red}] = \frac{\binom{5}{3} \times \binom{6}{1}}{330} = \frac{10 \times 6}{330} = \frac{60}{330}$$

$$P[4 \text{ red}] = \frac{\binom{5}{4}}{330} = \frac{5}{330}$$

Adding these probabilities, we obtain

$$P[\text{at least 2 red}] = \frac{150 + 60 + 5}{330} = \frac{215}{330} = 0.652$$

(c) $P[\text{all 4 red}] = \dfrac{\binom{5}{4}}{330} = \dfrac{5}{330}$

$$P[\text{all 4 yellow}] = \frac{\binom{6}{4}}{330} = \frac{15}{330}$$

Adding these probabilities, we obtain

$$P[\text{all 4 of the same color}] = \frac{20}{330} = 0.061.$$

4.103 (a) $S = \{1, 2, ..., 24\}$
        (b) $S = \{p : p > 0\}$, p is tire pressure in psi.
        (c) $S = \{0, 1, 2, ..., 50\}$
        (d) $S = \{t : t > 0\}$, t is time in days.

4.105 (a) $A = \{23, 24\}$
        (b) $A = \{p : 0 < p \le 28\}$
        (c) Since $0.25 \times 50 = 12.5$, $A = \{0, 1, ..., 12\}$
        (d) $A = \{t : 0 < t < 500.5\}$

4.107 $S = \{p : 0 \le p < 100\}$ where $p =$ percentage of alcohol in blood.

$A = \{p : .10 < p < 100\}$

4.109 The sample space is listed by means of a tree diagram which is given in Figure 4.4.

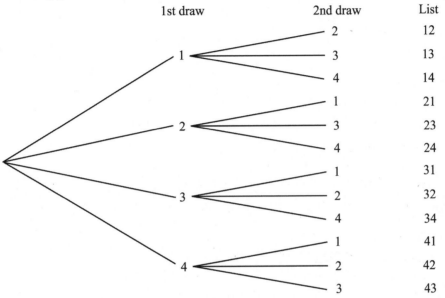

Figure 4.4: Sample space for Exercise 4.109

(a) S consists of 12 elementary outcomes which are equally likely because of random selection. Of the 12 elementary outcomes, 6 correspond to even numbers so

$$P[\text{even number}] = \frac{6}{12} = 0.5$$

(b) There are 9 elementary outcomes where the number is larger than 20, so the required probability is $\frac{9}{12} = 0.75$.

(c) There are 2 elementary outcomes, namely $\{23, 24\}$, for which the number is between 22 and 30, so the required probability is $\frac{2}{12} = 0.167$.

4.111   Consider the plot selected from each row to be assigned to variety 'a'. We list the
        sample space by drawing a tree diagram which is given in the figure below.

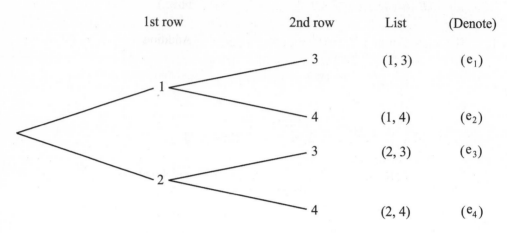

1st row	2nd row	List	(Denote)
	3	(1, 3)	(e_1)
	4	(1, 4)	(e_2)
	3	(2, 3)	(e_3)
	4	(2, 4)	(e_4)

$S = \{e_1, e_2, e_3, e_4\}$ and the elements are equally likely.

[same column] $= \{(1, 3), (2,4)\}$, Probability $= \dfrac{2}{4} = \dfrac{1}{2}$.

4.113   There are 12 letters which are equally likely to be selected.

(a) Since there are 5 vowels, the probability of getting a vowel is $\dfrac{5}{12}$.

(b) Among the 12 letters there are 3 T's, so the probability of getting a T is
$\dfrac{3}{12} = \dfrac{1}{4}$.

4.115   (a) $S = \{125, 152, 251, 215, 512, 521\}$

(b) $P(\text{less than 400}) = \dfrac{4}{6} = \dfrac{2}{3}$.

(c) $P(\text{even number}) = \dfrac{2}{6} = \dfrac{1}{3}$.

4.117   (a) Either a faulty transmission or faulty brakes.
        (b) Transmission, brakes and exhaust system all faulty.
        (c) No faults with the transmission, brakes or the exhaust system.
        (d) Either the transmission is not faulty or the brakes are not faulty.

4.119   Because $P(AB) \le P(A) \le P(A \cup B)$, we have $P(A) = 0.28$, $P(AB) = 0.1$, and
        $P(A \cup B) = 0.52$.

4.121 (a) ABC (b) $A \cup B \cup C$ (c) $AB\overline{C}$ (d) $\overline{A}B\overline{C}$

4.123 (a) $P(BC) = 0.05 + 0.20 = 0.25$ (see Figure 4.6 above.)

(b) $P(B \cup C) = P(B) + P(C) - P(BC)$ (Addition law)

$\qquad P(B) = 0.05 + 0.10 + 0.20 + 0.15 = 0.50$

$\qquad P(C) = 0.05 + 0.05 + 0.20 + 0.18 = 0.48$

$\quad P(BC) = 0.25$

Therefore, $P(B \cup C) = 0.50 + 0.48 - 0.25 = 0.73$
Alternatively,

$\qquad P(B \cup C) = P(B) + P(\overline{B}C)$ (See Figure 4.6 above.)

$\qquad\qquad = 0.50 + 0.23 = 0.73$

(c) $P(B\overline{C}) = 0.10 + 0.15 = 0.25$

(d) $P(A\overline{B}\overline{C} \cup \overline{A}B\overline{C} \cup \overline{A}\overline{B}C) = 0.17 + 0.15 + 0.18 = 0.50$

4.125 The probabilities can be determined either by using the Venn diagram or the probability table presented in the solution of Exercise 4.124.

(a) $P(B\overline{C}) = 0.05 + 0.20 = 0.25$
(by summing all probabilities in B but outside of C)

(b) $P(A \cup B) = P(A) + P(B) - P(AB)$ (Addition law)

$\qquad\qquad = 0.51 + 0.45 - 0.17$

$\qquad\qquad = 0.79$

(c) P[exactly two of the three events occur] $= P(AB\overline{C}) + P(A\overline{B}C) + P(\overline{A}BC)$

$\qquad\qquad\qquad\qquad = 0.05 + 0.21 + 0.08$ (see Figure 4.7)

$\qquad\qquad\qquad\qquad = 0.34$.

4.127 $P(A \mid B) = \dfrac{P(AB)}{P(B)} = \dfrac{0.3}{0.5} = 0.6$

$P(A)P(B) = 0.6 \times 0.5 = 0.30 = P(AB)$, so A and B are independent.

4.129 (a) $P(AC) = 0.15$

$\qquad P(A)P(C) = 0.6 \times 0.25 = 0.15$

Because $P(AC) = P(A)P(C)$, the events A and C are independent.

(b) $P(A\overline{B}C) = P(AC) = 0.15$ (see the Venn diagram in Exercise 4.128)

$\qquad P(A\overline{B}) = 0.25 + 0.15 = 0.40$

$P(C) = 0.15 + 0.1 = 0.25$

$P(A\overline{B})P(C) = 0.40 \times 0.25 = 0.1 \neq P(A\overline{B}C)$, so the events $A\overline{B}$ and C are not independent.

4.131 (a) $P(A \mid \overline{B}) = \dfrac{P(A\overline{B})}{P(\overline{B})}$

From the probability table in Exercise 4.122, we get

$P(A\overline{B}) = 0.05 + 0.17 = 0.22$

$P(\overline{B}) = 0.05 + 0.17 + 0.18 + 0.10 = 0.50$

So, $P(A \mid \overline{B}) = \dfrac{0.22}{0.50} = 0.44$.

(b) $P(B \mid AC) = \dfrac{P(ABC)}{P(AC)} = \dfrac{0.05}{0.10} = 0.50$

(c) From the probability table, we find $P(A) = 0.37$, $P(C) = 0.48$. $P(AC) = 0.10$. Since $P(A)P(C) = 0.37 \times 0.48 = 0.178 \neq P(AC)$, the events A and C are dependent.

4.133 (a) $S = \{e_1, e_2, e_3, e_4\}$ where e_1 corresponds to the selection sequence 11, e_2 to 12, e_3 to 21, and e_4 to 22.

$P(e_1) = \dfrac{10}{15} \times \dfrac{9}{14} = \dfrac{90}{210}$ (Because the probability of a #1 marble in the first

draw is $\dfrac{10}{15}$, and the conditional probability of a #1 marble in the second draw

given that a #1 marble appears in the first draw, is $\dfrac{9}{14}$.)

Similarly, we calculate

$$P(e_2) = \dfrac{10}{15} \times \dfrac{5}{14} = \dfrac{50}{210}$$
$$P(e_3) = \dfrac{5}{15} \times \dfrac{10}{14} = \dfrac{50}{210}$$
$$P(e_4) = \dfrac{5}{15} \times \dfrac{4}{14} = \dfrac{20}{210}$$

(b) [Even number] $= \{e_2, e_4\}$, Probability $= \dfrac{50}{210} + \dfrac{20}{210} = \dfrac{70}{210} = \dfrac{1}{3}$

(c) [Larger than 15] $= \{e_3, e_4\}$, Probability $= \dfrac{50}{210} + \dfrac{20}{210} = \dfrac{70}{210} = \dfrac{1}{3}$.

4.135 (a) & (b) The tree diagram is given below.

(i) $P(FF) + P(FD) + P(FN) + P(DF) + P(NF) = 0.64$

(ii) $P(FF) + P(FD) + P(DF) + P(DD) = 0.49$

Subject 1	Subject 2	List	Probability
	F	FF	$(0.4)(0.4) = 0.16$
F	D	FD	$(0.4)(0.3) = 0.12$
	N	FN	$(0.4)(0.3) = 0.12$
	F	DF	$(0.3)(0.4) = 0.12$
D	D	DD	$(0.3)(0.3) = 0.09$
	N	DN	$(0.3)(0.3) = 0.09$
	F	IF	$(0.3)(0.4) = 0.12$
N	D	ID	$(0.3)(0.3) = 0.09$
	N	IN	$(0.3)(0.3) = 0.09$
			$\overline{}$
			1.00

4.137 There are $\binom{11}{3} = \dfrac{11 \times 10 \times 9}{3 \times 2 \times 1} = 165$ ways to select three trucks at random. If the company's argument is correct, that there are exactly three noncompliant trucks, then only one selection can consist of three noncompliant vehicles. The probability of selecting all three noncompliant trucks is $\dfrac{1}{165} = 0.0061$, a small probability. It is quite unlikely a random selection would produce the indicated sample so we can question the veracity of the company's claim.

4.139 Denote G = Good standing, D = Illegal deduction.

$$\begin{array}{|cc|} 11 & G \\ 7 & D \\ \hline \end{array} \ \rightarrow 4$$
$$18$$

(a) The number of possible selections of 4 returns out of 18 is

$$\binom{18}{4} = 3060, \ \text{all equally likely.}$$

$$P[\text{Sample has all 4 } G\text{'s}] = \frac{\binom{11}{4}}{\binom{18}{4}} = \frac{330}{3060} = .108 \,.$$

(b) $P[\text{at least 2 } D\text{'s}] = P[2D's] + P[3D's] + P[4D's]$

$$P[2D's] = \frac{\binom{7}{2}\binom{11}{2}}{\binom{18}{4}} = \frac{21 \times 55}{3060} = \frac{1155}{3060}$$

$$P[3D's] = \frac{\binom{7}{3}\binom{11}{1}}{\binom{18}{4}} = \frac{35 \times 11}{3060} = \frac{385}{3060}$$

$$P[4D's] = \frac{\binom{7}{4}}{\binom{18}{4}} = \frac{35}{3060}$$

Adding these we obtain $P[\text{at least 2 } D\text{'s}] = \dfrac{1575}{3060} = 0.515 \,.$

4.141 (a) $P(\text{no common birthday}) = \dfrac{365 \times 364 \times 363}{365 \times 365 \times 365} = 1 \times \dfrac{364}{365} \times \dfrac{363}{365} = 0.992 \,.$

Hence, $P(\text{at least two have the same birthday}) = 1 - 0.992 = 0.008 \,.$

(b) For each person there are 365 possible birthdays. Hence, for N persons, the number of possible birthdays is

$$365 \times 365 \times \ldots \times 365 \ (N \text{ factors}) = (365)^N \,.$$

In order that the birthdays of N persons to be all different, the first person can have any of the 365 days, the second person can have any of the remaining $365 - 1 = 364$ days, the third person can have any of the remaining $365 - 2 = 363$ days, ... the Nth person can have any of the remaining $365 - N + 1$ days.

Hence, the number of ways N persons can have all different birthdays is $365 \times 364 \times \ldots \times (365 - N + 1) \,.$

Chapter 5

PROBABILITY DISTRIBUTIONS

5.1 (a) Discrete,   (b) Continuous,   (c) Continuous

(d) Continuous,   (e) Discrete

5.3 (a)

Possible choices	x
{1, 3}	2
{1, 5}	4
{1, 6}	5
{1, 7}	6
{3, 5}	2

Possible choices	x
{3, 6}	3
{3, 7}	4
{5, 6}	1
{5, 7}	2
{6, 7}	1

(b) The distinct values of x and the corresponding probabilities are listed in the following table. All 10 choices, listed in part (a), are equally likely so

$$P[X = x] = \frac{\text{No. Choices for which } X = x}{10}$$

x	$P[X = x]$
1	0.2
2	0.3
3	0.1
4	0.2
5	0.1
6	0.1

5.5   (a) The ratings by the judges are given in the figure below.

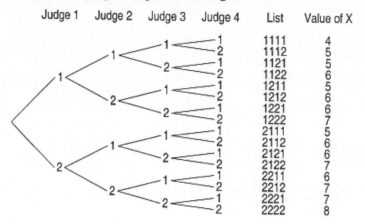

(b) 4, 5, 6, 7, 8

5.7   (a) & (b) The outcomes and the values of X are given on the tree diagram below.

Week 1	Week 2	Week 3	List	Value of X
		C	CCC	0
	C			
		B	CCB	1
C		C	CBC	2
	B			
		B	CBB	1
		C	BCC	1
	C			
		B	BCB	2
B		C	BBC	1
	B			
		B	BBB	0

5.9   The distinct values of X are 0, 2, 4. Since $[X = 0] = \{e_2, e_4, e_6, e_7\}$, we obtain

$$f(0) = P[X = 0] = P(e_2) + P(e_4) + P(e_6) + P(e_7)$$
$$= 0.30 + 0.10 + 0.11 + 0.13 = 0.64$$

Similarly, we calculate

$$f(2) = P[X = 2] = 0.05 + 0.15 = 0.20$$
$$f(4) = P[X = 4] = 0.16$$

The probability distribution of X is given in the following table.

x	$P[X = x]$
0	0.64
2	0.20
4	0.16
Total	1.00

5.11 (a) Possible values of X are 2, 3, 4, 5, 6, 7, 8, 9, 10, 11, 12.

(b)
$$X = 2 \quad (1,1)$$
$$X = 3 \quad (1,2), \quad (2,1)$$
$$X = 4 \quad (1,3), \quad (2,2), \quad (3,1)$$
$$X = 5 \quad (1,4), \quad (2,3), \quad (3,2), \quad (4,1)$$
$$X = 6 \quad (1,5), \quad (2,4), \quad (3,3), \quad (4,2), \quad (5,1)$$
$$X = 7 \quad (1,6), \quad (2,5), \quad (3,4), \quad (4,3), \quad (5,2), \quad (6,1)$$
$$X = 8 \quad (2,6), \quad (3,5), \quad (4,4), \quad (5,3), \quad (6,2)$$
$$X = 9 \quad (3,6), \quad (4,5), \quad (5,4), \quad (6,3)$$
$$X = 10 \quad (4,6), \quad (5,5), \quad (6,4)$$
$$X = 11 \quad (5,6), \quad (6,5)$$
$$X = 12 \quad (6,6)$$

(c) The 36 possible outcomes are equally likely so $P[X = x]$ is the number of outcomes for which $X = x$ divided by 36. The probability distribution of X is

x	$f(x)$
2	1/36
3	2/36
4	3/36
5	4/36
6	5/36
7	6/36
8	5/36
9	5/36
10	4/36
11	2/36
12	1/36
Total	1

5.13 (a)

x	$f(x)$
2	1/10
3	2/10
4	3/10
5	4/10
Total	1

Yes, a probability distribution.

(b)

x	$f(x)$
1	−1/3
2	0
3	1/3
4	2/3
Total	2/3

Not a probability distribution because $f(1)$ is negative.

(c)

x	$f(x)$
−2	0
−1	2/20
0	4/20
1	6/20
2	8/20
Total	1

(d)

x	$f(x)$
2	3/4
3	3/8
4	3/16
Total	21/16

Yes, a probability distribution.   No, because $\sum f(x)$ does not equal 1.

5.15 Since $P(B) = \dfrac{1}{2}$, $P(C) = \dfrac{1}{2}$ and purchases in different weeks are independent, the probability model is the same as that for three tosses of a fair coin. Each elementary outcome has probability $\dfrac{1}{8}$. Therefore

$$P[X = 0] = \frac{2}{8}, \quad P[X = 1] = \frac{4}{8}, \quad P[X = 2] = \frac{2}{8}.$$

The probability distribution of X is

x	$f(x)$
0	1/4
1	1/2
2	1/4

5.17 The selected card can have any of the numbers −3, −1, 0, 1, 3 with equal probabilities $1/5$. Define the random variable X as follows:

X = the square of the selected number, can assume the values 0, 1, 9. Then,

$[X = 0] = [\text{The selected number} = 0]$, $P[X = 0] = \dfrac{1}{5}$

$[X = 1] = [\text{The selected number} = -1 \text{ or } 1]$, $P[X = 1] = \dfrac{2}{5}$

$[X = 9] = [\text{The selected number} = -3 \text{ or } 3]$, $P[X = 9] = \dfrac{2}{5}$

The probability distribution of X is

x	$f(x)$
0	1/5
1	2/5
9	2/5
Total	1

5.19 (a) Since $3+1+1+1+1+3=10$, the probabilities of the six faces numbered 1, 2, 3,

4, 5, 6 are $\dfrac{3}{10}, \dfrac{1}{10}, \dfrac{1}{10}, \dfrac{1}{10}, \dfrac{1}{10}, \dfrac{3}{10}$, respectively. The probability distribution

of X is

x	$f(x)$
1	0.3
2	0.1
3	0.1
4	0.1
5	0.1
6	0.3
Total	1

(b) $P[\text{Even number}] = f(2) + f(4) + f(6) = 0.1 + 0.1 + 0.3 = 0.5$.

5.21 Since $P[X \text{ is odd}] = f(1) + f(3) + f(5) = 0.1 + 0 + 0.3 = 0.4$ we have
$P[X \text{ is even}] = 1 - .4 = .6$, that is, $f(2) + f(4) + f(6) = 0.6$.   Now, in order that
$f(2)$, $f(4)$ and $f(6)$ are all equal, and their total is 0.6, we must have
$f(2) = 0.2$, $f(4) = 0.2$, $f(6) = 0.2$. Thus, the probability distribution of X is given
below.

x	$f(x)$
1	0.1
2	0.2
3	0
4	0.2
5	0.3
6	0.2
Total	1

5.23 (a) Consider random selection of one ball from an urn that contains the following
mix of 100 numbered balls: 32 balls are numbered 2, 44 balls are numbered 4,
and 24 balls are numbered 6. If X denotes the number on the selected ball, then
the probability distribution of X would be as given in Table (a).

(b) Consider random selection of one ball from an urn that contains the following
mix of 14 numbered balls: 3 balls are numbered -2, 4 balls are numbered 0, 5
balls are numbered 4, and 2 balls are numbered 5. If X denotes the number on
the selected ball, then the probability distribution of X would be as given in
Table (b).

5.25 (a) $P[X \le 3] = f(0) + f(1) + f(2) + f(3) = 0.90$
      (b) $P[X \ge 2] = f(2) + f(3) + f(4) = 0.63$
      (c) $P[1 \le X \le 3] = f(1) + f(2) + f(3) = 0.78$.

5.27 Here X = the number of customers per day.

    (a) Customers will be turned away if there are 3 or more customers. The required
        probability is
$$P[X \ge 3] = f(3) + f(4) + f(5) = 0.25 + 0.15 + 0.05 = 0.45.$$
    (b) The center's capacity is not fully utilized if fewer than 2 customers arrive. This
        probability is
$$P[X \le 1] = f(0) + f(1) = 0.05 + 0.20 = 0.25.$$
    (c) We see that
$$P[X \ge 5] = 0.05 \text{ and } P[X \ge 4] = 0.15 + 0.05 = 0.20 > 0.10.$$
    Therefore, the capacity must be increased by 2. With a capacity of 4, the
    probability of turning customers away is $P[X \ge 5] = 0.05$.

5.29 (a) The probability histogram of X is given below.

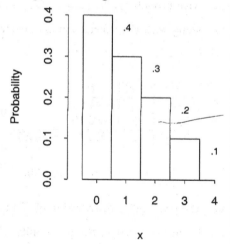

    (b) The calculation is given in the following table:

x	$f(x)$	$xf(x)$	$(x-\mu)$	$(x-\mu)^2 f(x)$
1	0.4	0.4	−1	0.4
2	0.3	0.6	0	0
3	0.2	0.6	1	0.2
4	0.1	0.4	2	0.4
Total		$2.0 = \mu$		$1.0 = \sigma^2$

$$E(X) = 2, \quad \sigma^2 = 1, \quad \sigma = 1$$

5.31 Let X denote the carpenter's net profit. Then $X = \$5,000$ with probability 0.2 and
$X = -\$56$ (loss) with probability $1 - 0.2 = 0.8$. The probability distribution is
shown in the table along with the calculation of expectation.

x	$f(x)$	$xf(x)$
-56	0.8	-44.80
$5,000$	0.2	1000.00
Total	1	$955.20 = E(X)$

So, the expected return $= \$955.20$.

5.33

x	$f(x)$	$xf(x)$	$x^2 f(x)$
0	0.2401	0	0
1	0.4116	0.4116	0.4116
2	0.2646	0.5292	1.0584
3	0.0756	0.2268	0.6804
4	0.0081	0.0324	0.1296
Total		1.2000	2.2800

$$\mu = 1.2$$
$$\sigma^2 = 2.28 - (1.2)^2 = 0.84, \quad \sigma = \sqrt{0.84} = 0.917$$

5.35 (a) We denote "win" by W and "not win" by N, and attach suffixes A or B to
identify the project. Listed here are the possible outcomes and calculation of
the corresponding probabilities. For instance,

$$P(W_A N_B) = P(W_A)P(N_B), \text{ by independence}$$
$$= 0.50 \times 0.35 = 0.175$$

Outcome	Probability
$W_A W_B$	$0.50 \times 0.65 = 0.325$
$W_A N_B$	$0.50 \times 0.35 = 0.175$
$N_A W_B$	$0.50 \times 0.65 = 0.325$
$N_A N_B$	$0.50 \times 0.35 = 0.175$

(b) & (c) The amounts of profit (X) for the various outcomes are listed below.

Outcome	Profit (\$) X
$W_A W_B$	$75,000 + 120,000 = 195,000$
$W_A N_B$	$75,000$
$N_A W_B$	$120,000$
$N_A N_B$	0

In the next table, we present the probability distribution of X and calculate $E(X)$.

x	$f(x)$	$xf(x)$
0	0.175	0
$75,000$	0.175	$13,125$
$120,000$	0.325	$3,900$
$195,000$	0.325	$63,375$
Total		$80,400 \;\; = E(X)$

Expected net profit $= E(X) - \text{cost} = \$80,400 - \$2,000 = \$78,400$.

5.37 (a) & (b) The expectation and standard deviation of X are calculated in the following table.

x	$f(x)$	$xf(x)$	$x^2 f(x)$
0	0.315	0	0
1	0.289	0.289	0.289
2	0.201	0.402	0.804
3	0.114	0.342	1.026
4	0.063	0.252	1.008
5	0.012	0.060	0.300
6	0.006	0.036	0.216
Total		1.381	3.643

$E(X)$ or $\mu = 1.381$

$\text{Var}(X)$ or $\sigma^2 = 3.643 - (1.381)^2 = 1.736$

Standard deviation of X is $\sigma = \sqrt{1.736} = 1.318$.

5.39 (a) To tabulate the probability distribution, we calculate the function $f(x)$ for $x = 0, 1, 2, 3$.

$$f(0) = \frac{1}{84} \binom{5}{0} \binom{4}{3} = \frac{4}{84}$$

$$f(1) = \frac{1}{84} \binom{5}{1} \binom{4}{2} = \frac{5 \times 6}{84} = \frac{30}{84}$$

$$f(2) = \frac{1}{84} \binom{5}{2} \binom{4}{1} = \frac{10 \times 4}{84} = \frac{40}{84}$$

$$f(3) = \frac{1}{84} \binom{5}{3} \binom{4}{0} = \frac{10}{84}$$

x	$f(x)$	$xf(x)$	$x^2 f(x)$
0	4/84	0	0
1	30/84	30/84	30/84
2	40/84	80/84	160/84
3	10/84	30/84	90/84
Total	1	140/84	280/84

(b) Referring to the calculations shown in the table in part (a), we find

$$\text{mean} = \sum xf(x) = \frac{140}{84} = 1.667$$

$$\text{variance} \quad \sigma^2 = \sum x^2 f(x) - \left[\sum xf(x)\right]^2$$

$$= \frac{280}{84} - \left(\frac{140}{84}\right)^2 = 0.556$$

$$\text{standard deviation} \quad \sigma = \sqrt{0.556} = 0.745.$$

5.41 (a) & (b) Let us use the symbol S for a sale and N for no sale at a customer contact. For contact with 4 customers, we list the elementary outcomes. In assigning the probabilities we use the assumption of independence and the facts that $P(S) = 0.3$, $P(N) = 0.7$. For instance,

$$P(SNNS) = 0.3 \times 0.7 \times 0.7 \times 0.3 = 0.0441.$$

Elementary outcome	Probability	Value of X
NNNN	$(0.7)^4 = 0.2401$	0
NNNS	$(0.7)^3 \times (0.3) = 0.1029$	1
NNSN	0.1029	1
NSNN	0.1029	1
SNNN	0.1029	1
NNSS	$(0.7)^2 \times (0.3)^2 = 0.0441$	2
NSNS	0.0441	2
SNNS	0.0441	2
NSSN	0.0441	2
SNSN	0.0441	2
SSNN	0.0441	2
NSSS	$0.7 \times (0.3)^3 = 0.0189$	3
SNSS	0.0189	3
SSNS	0.0189	3
SSSN	0.0189	3
SSSS	$(0.3)^4 = \underline{0.0081}$	4
	Total 1.0000	

$$P[X = 0] = 0.2401, \qquad P[X = 1] = 4 \times 0.1029 = 0.4116$$
$$P[X = 2] = 6 \times 0.0441 = 0.2646, \quad P[X = 3] = 4 \times 0.0189 = 0.0756$$
$$P[X = 4] = 0.0081$$

Probability distribution of X and calculation of expectation

x	$f(x)$	$xf(x)$
0	0.2401	0
1	0.4116	0.4116
2	0.2646	0.5292
3	0.0756	0.2268
4	0.0081	0.0324
Total	1.0000	1.2000

(c) $E(X) = 1.2$.

5.43 $P[X \leq 2] = 0.4 + 0.3 = 0.7$ and $P[X \geq 2] = 0.3 + 0.2 + 0.1 = 0.6$
Since both are ≥ 0.5, median $= 2$.

5.45 (a) The Bernoulli model is not appropriate. The assumption of independence is likely to be violated because of peer pressure.

(b) The Bernoulli model is not appropriate because the measurement is on a continuous scale.

(c) The Bernoulli model may be appropriate because there are only two possible outcomes. Independence may be violated for items close together.

(d) Here, each house is a trial with the two possible outcomes: the delivery was on time or was not. However, the Bernoulli model does not seem plausible because of a lack of independence of the trials.

5.47 Since the event of interest is the drawing of a yellow candy, we identify $S =$ yellow, $F =$ any color other than yellow.

(a) Because the sampling is with replacement, and each draw has two possible outcomes S or F, the model of Bernoulli trials is appropriate. The mix has 25 candies of which 10 are yellow so the probability of drawing a yellow candy is $\dfrac{10}{25} = 0.4$. We have $p = 0.4$.

(b) The model of Bernoulli trials is not appropriate because the sampling is without replacement and the size of the lot is not large. The condition of independence of the trials is violated.

(c) The condition of independence of outcomes in the different trials is violated. For instance, $P(S_2 \mid S_1) = \dfrac{11}{26}$ while $P(S_2 \mid F_1) = \dfrac{10}{26}$. The model of Bernoulli trials is not appropriate.

5.49 Label the plots 1, 2, 3 and 4. The number of possible selections of 2 plots out of 4 is $\dbinom{4}{2} = 6$, and these are equally likely.

(a) Consider the event of an S in the first trial, that is, the first plot is selected. One other plot can be chosen from plots 2, 3, and 4 in $\dbinom{3}{1} = 3$ ways. Therefore,

$$P(S \text{ in first trial}) = \frac{3}{6} = \frac{1}{2}.$$

The same argument leads to $P(S) = \dfrac{1}{2}$ in any particular trial.

(b) Denote $S_1S_2 =$ the event that the first and second plots are selected. We have $P(S_1S_2) = \dfrac{1}{6}$. From part (a), we find that

$$P(S_1)P(S_2) = \frac{1}{2} \times \frac{1}{2} = \frac{1}{4} \neq P(S_1S_2)$$

Therefore, the trials are not independent.

5.51 (a) Although there are two possible outcomes of each trial, the Bernoulli model is not appropriate because the 5 purchases of each consumer cannot be considered independent.

(b) Here the Bernoulli model is plausible because the 500 trials correspond to different consumers who are selected at random.

5.53 We have $P(S) = p = \dfrac{1}{3}$, $P(F) = q = 1 - \dfrac{1}{3} = \dfrac{2}{3}$.

(a) $P(FFFF) = \dfrac{2}{3} \times \dfrac{2}{3} \times \dfrac{2}{3} \times \dfrac{2}{3} = \dfrac{16}{81} = 0.1975$

(b) Because the trials are independent, the required conditional probability is the same as the (unconditional) probability of 4 trials resulting in all successes, which is

$$P(SSSS) = \left(\frac{1}{3}\right)^4 = \frac{1}{81} = 0.0123$$

(c) $P(FFFS) = \dfrac{2}{3} \times \dfrac{2}{3} \times \dfrac{2}{3} \times \dfrac{1}{3} = \dfrac{8}{81} = 0.0988$

5.55 (a) $P(SSSS) = 0.8 \times 0.8 \times 0.8 \times 0.8 = 0.4096$

(b) For a single trial $P(F) = 1 - 0.8 = 0.2$ so $P(FFFF) = (0.2)^4 = 0.0016$

(c) Using the law of complement,

$$P[\text{at least one success}] = 1 - P[\text{no successes}] = 1 - 0.0016 \quad (\text{see (b)})$$
$$= 0.9984.$$

5.57 (a) The possible results in the first two trials are SS, SD, DS, DD. If SS occurs, the experiment is stopped. With SD, there is one more trial so we have either SDS or SDD. Proceeding in this way, the complete list is

$$\{SS, SDS, SDD, DSS, DSD, DDSS, DDSD, DDDS, DDDD\}$$

(b) We have $P(S) = 1/4$, $P(D) = 3/4$

Outcome	Probability	Value of X
SS	$(1/4)^2 = 1/16$	2
SDS	$(1/4)^2(3/4) = 3/64$	2
SDD	$9/64$	1
DSS	$3/64$	2
DSD	$9/64$	1
DDSS	$9/256$	2
DDSD	$27/256$	1
DDDS	$27/256$	1
DDDD	$81/256$	0

(c) The probability distribution of X is:

x	$f(x)$
0	$81/256$
1	$126/256$
2	$49/256$
Total	1

5.59 (a) Yes. $n = 10$, $p = \dfrac{1}{6}$

(b) No, because the number of trials is not fixed.

(c) Yes. $n = 3$, and p is the probability of getting a marble numbered either 1 or 2 in a single draw so
$$p = \frac{4+3}{10} = 0.7.$$

(d) No, because X does not represent a count of the number of times that an event occurs.

5.61 (a) With $n = 3$, $p = 0.35$, $q = 0.65$ we obtain
$$P[X = 2] = \binom{3}{2}(0.35)^2(0.65) = 0.2389$$

(b) $n = 6$, $p = 0.25$, $q = 0.75$
$$P[X = 3] = \binom{6}{3}(0.25)^3(0.75)^3 = 0.132$$

(c) $n = 6$, $p = 0.75$, $q = 0.25$
$$P[X = 2] = \binom{6}{2}(0.75)^2(0.25)^4 = 0.033$$

5.63 $n = 5$, $p = 0.35$

$$f(x) = \binom{5}{x}(0.35)^x(0.65)^{5-x}$$

The calculation of $f(x)$ is presented in the following table:

x	$f(x)$	
0	$(0.65)^5$	$= 0.1160$
1	$5(0.35)(0.65)^4$	$= 0.3124$
2	$10(0.35)^2(0.65)^3$	$= 0.3364$
3	$10(0.35)^3(0.65)^2$	$= 0.1811$
4	$5(0.35)^4(0.65)$	$= 0.0488$
5	$(0.35)^5$	$= 0.0053$
	Total	1.0000

(a) $P[X \leq 3] = f(0) + f(1) + f(2) + f(3) = 0.9459$
(b) $P[X \geq 3] = f(3) + f(4) + f(5) = 0.2352$
(c) $P[X = 2 \text{ or } 4] = f(2) + f(4) = 0.3852$

5.65 $n = 4$, $p = 0.45$, $q = 0.55$

$$f(x) = \binom{4}{x}(0.45)^x(0.55)^{4-x}$$

To calculate the required probabilities, it would be convenient to calculate all the $f(x)$ values.

x	$f(x)$	
0	$(0.55)^4$	$= 0.0915$
1	$4(0.45)(0.55)^3$	$= 0.2995$
2	$6(0.45)^2(0.55)^2$	$= 0.3675$
3	$4(0.45)^3(0.55)$	$= 0.2005$
4	$(0.45)^4$	$= 0.0410$
	Total	1.0000

(a) $P[X \geq 3] = f(3) + f(4) = 0.2415$
(b) $P[X \leq 3] = 1 - f(4) = 0.9590$
(c) "Two or more failures" means two or fewer successes. So, the required probability is
$$P[X \leq 2] = f(0) + f(1) + f(2) = 0.7585$$

5.67 Identify S : severe leaf damage.

$X = $ Number of trees with severe leaf damage in a random sample of 5 trees.

X has a binomial distribution with $n = 5$, $p = 0.15$, $q = 0.85$.

$$f(x) = \binom{5}{x}(0.15)^x (0.85)^{5-x}, \quad x = 0, 1, ..., 5$$

(a) $P[X = 3] = f(3) = \binom{5}{3}(0.15)^3 (0.85)^2 = 0.024$

(b) $P[X \leq 2] = f(0) + f(1) + f(2) = 0.974$.

5.69 (a) We use the binomial table for $n = 13$, $p = 0.3$.
$$P[X = 4] = P[X \leq 4] - P[X \leq 3] = 0.654 - 0.421 = 0.233$$

(b) '8 failures in 13 trials' means 5 successes in 13 trials. We use the binomial table for $n = 13$, $p = 0.7$
$$P[X = 5] = 0.018 - 0.004 = 0.014$$

(c) Using the binomial table for $n = 13$, $p = 0.3$, we obtain
$$P[X = 8] = 0.996 - 0.982 = 0.014$$

Refer to (b), and consider interchanging the names 'success' and 'failure'. Then the new p would be .3, (i.e., the old q), and the specified event would then be '8 successes in 13 trials' which is precisely the statement in part (c).

5.71 (a) Denote $X = $ number of successes in 5 trials.

The event 'more than 5 trials are needed in order to obtain 3 successes' means that at most two successes are obtained in 5 trials, that is, $X \leq 2$. Since X has the binomial distribution with $n = 5$, $p = 0.7$, we use the binomial table to find the required probability
$$P[X \leq 2] = 0.163.$$

(b) The stated event is equivalent to 'at most 6 successes in 9 trials', that is, $X \leq 6$ where X denotes the number of successes in 9 trials. Using the binomial table with $n = 9$, $p = 0.7$, we find the required probability
$$P[X \leq 6] = 0.537$$

5.73 Mean $= np$, $sd = \sqrt{npq}$

(a) Mean $= 19 \times 0.5 = 9.5$, $\quad sd = \sqrt{19 \times 0.5 \times 0.5} = 2.179$

(b) Mean $= 25 \times 0.2 = 5$, $\quad sd = \sqrt{25 \times 0.2 \times 0.8} = 2.0$

(c) Mean $= 25 \times 0.8 = 20$, $\quad sd = \sqrt{25 \times 0.2 \times 0.8} = 2.0$

5.75 Denote X = number of college seniors in support of increased funding, in a random sample of 20 seniors.

Then, X has the binomial distribution with $n = 20$, $p = 0.2$. We find
$P[X \leq 3] = 0.411$.

5.77 Identify S : survival beyond 5 years

Denote X = number of patients surviving beyond 5 years, in a random sample of 19 patients.

Then, X has the binomial distribution with $n = 19$, $p = 0.8$

(a) $P[X = 14] = 0.327 - 0.163 = 0.164$
(b) $P[X = 19 - 6] = P[X = 13] = 0.163 - 0.068 = 0.095$
(c) $P[9 \leq X \leq 13] = P[X \leq 13] - P[X \leq 8] = 0.163 - 0.000 = 0.163$

5.79 $n = 545$, $p = 0.0785$, $q = 0.9215$
mean $= np = 42.7825$
$sd = \sqrt{npq} = 6.2789$.

5.81 (a) X has the binomial distribution with $n = 40$ and

$$p = P[\text{Allergy present}] = 0.16 + 0.09 = 0.25$$

Therefore,
$$E(X) = 40 \times 0.25 = 10$$
$$sd(X) = \sqrt{40 \times 0.25 \times 0.75} = 2.739$$

(b) Y has the binomial distribution with $n = 40$ and

$$p = P[\text{Allergy present - Male}] = \frac{0.16}{0.16 + 0.36} = \frac{16}{52}$$

Therefore,
$$E(Y) = 40 \times \frac{16}{52} = 12.308$$
$$sd(Y) = \sqrt{40 \times \frac{16}{52} \times \frac{36}{52}} = 2.919$$

(c) Z has the binomial distribution with $n = 40$ and

$$p = P[\text{Allergy absent - Female}] = \frac{0.39}{0.09 + 0.39} = \frac{39}{48}$$

Therefore,

$$E(Z) = 40 \times \frac{39}{48} = 3.25$$

$$sd(Z) = \sqrt{40 \times \frac{39}{48} \times \frac{9}{48}} = 2.469$$

5.83 Minitab binomial $n = 12$, $p = 0.67$

(a) $P[X \le 8] = 0.5973$
     $P[X = 8] = 0.2384$
(b) $n = 35$, $p = 0.43$, $P[10 \le X \le 15] = 0.5372$

5.85 (a) From the example, the center line is at $p_0 = 0.4$, the lower control limit is 0.07 and the upper control limit is 0.73.

(b) The corresponding proportions 0.6, 0.5, 0.75, 0.55, and 0.8 are graphed below.

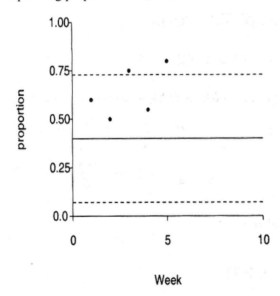

(c) The proportions 0.75 and 0.8, for the third and fifth weeks, are out of control.

5.87 (a) From Exercise 5.86, the lower limit is 0.165 and the upper limit is 0.835. Since,

$$\frac{x}{20} \le 0.165 \quad \text{only if} \quad x \le 20 \times 0.165 = 3.3$$

and

$$\frac{x}{20} \geq 0.165 \quad \text{only if} \quad x \geq 20 \times 0.835 = 16.7$$

we see that the unusual values are 0, 1, 2, 3 and 17, 18, 19, 20.

(b) When $n = 20$, $p = 0.5$,
$$P[X \leq 3 \quad or \quad X \geq 17] = P[X \leq 3] + (1 - P[X \leq 16]) = 0.001 + (1 - 0.999) = 0.002$$

5.89 (a) Possible values of X are $-3, -1, 1, 3$.

(b) $X = -3$: TTT
$X = -1$: HTT, THT, TTH
$X = 1$: HHT, HTH, THH
$X = 3$: HHH

5.91 (a) Considering the selection of the box and the possible number of defectives in the sample, the elementary outcomes are

$$(\text{Box } 1, \ 0D), \quad (\text{Box } 2, \ 0D)$$
$$(\text{Box } 1, \ 1D), \quad (\text{Box } 2, \ 1D)$$
$$(\text{Box } 1, \ 2D), \quad (\text{Box } 2, \ 2D)$$

Using conditional probability and the multiplication law, we calculate

$$P[\text{Box } 1, \ 0D] = P(\text{Box } 1)P(0D \,|\, \text{Box } 1)$$

$$= \frac{1}{2} \times \frac{\binom{15}{2}}{\binom{20}{2}}$$

because Box 1 contains 20 articles of which 15 are good and 5 are defective

$$= \frac{1}{2} \times \frac{105}{190} = 0.2763$$

In the same manner we obtain

$$P[\text{Box } 1, \ 1D] = \frac{1}{2} \times \frac{\binom{5}{1}\binom{15}{1}}{\binom{20}{2}} = \frac{1}{2} \times \frac{75}{190} = 0.1974$$

$$P[\text{Box } 1, \ 2D] = \frac{1}{2} \times \frac{\binom{5}{2}}{\binom{20}{2}} = \frac{1}{2} \times \frac{10}{190} = 0.0263$$

$$P[\text{Box 2, } 0D] = \frac{1}{2} \times \frac{\dbinom{24}{2}}{\dbinom{30}{2}} = \frac{1}{2} \times \frac{276}{435} = 0.3172$$

$$P[\text{Box 2, } 1D] = \frac{1}{2} \times \frac{\dbinom{6}{1}\dbinom{24}{1}}{\dbinom{30}{2}} = \frac{1}{2} \times \frac{144}{435} = 0.1655$$

$$P[\text{Box 2, } 2D] = \frac{1}{2} \times \frac{\dbinom{6}{2}}{\dbinom{30}{2}} = \frac{1}{2} \times \frac{15}{435} = 0.0172$$

(b) The random variable Y, the number of defectives in the sample, can have the values 0, 1, 2, and the probabilities are readily obtained by adding the probabilities of the relevant elementary outcomes. For instance,

$$P[Y = 0] = P[\text{Box 1, } 0D] + P[\text{Box 2, } 0D] = 0.2763 + 0.3172 = 0.5935$$

The probability distribution of Y is

y	$f(y)$
0	0.5935
1	0.3629
2	0.0435
Total	0.9999 (rounding error)

5.93 (a) We denote the results of the successive matches by a string of letters A and B indicating the winner of each individual set. For instance, $AABA$ stands for the outcome that A wins the first two sets, B wins the third set and A wins the fourth. In this case, the game stops at the fourth set so $X = 4$. Listed below are the possible values of X and the corresponding elementary outcomes:

$X = 3:$ $AAA,$

        BBB

$X = 4:$ $AABA, ABAA, BAAA,$

        $BBAB, BABB, ABBB$

$X = 5:$ $AABBA, ABABA, ABBAA, BAABA, BABAA, BBAAA,$

        $BBAAB, BABAB, BAABB, ABBAB, ABABB, AABBB$

(Note: In each case, the outcomes in the first line correspond to A being the winner and those in the second line correspond to B being the winner. Once the first line is completed, the second line can be readily obtained by interchanging the letters A and B).

(b) To calculate the probabilities we use the facts that $P(A) = 0.4$, $P(B) = 0.6$, and the results of the different sets are independent.

$$P[X = 3] = (0.4)^3 + (0.6)^3 = 0.280$$
$$P[X = 4] = 3(0.4)^3(0.6) + 3(0.6)^3(0.4) = 0.3744$$
$$P[X = 5] = 6(0.4)^3(0.6)^2 + 6(0.6)^3(0.4)^2 = 0.3456$$

The probability distribution of X is

x	$f(x)$
3	0.2800
4	0.3744
5	0.3456
Total	1.0000

5.95

$$
\begin{array}{c|c}
4 & \text{defective} \\
8 & \text{good} \\
\hline
12 &
\end{array}
\;\rightarrow\; 3 \text{ sampled}
$$

The possible values of X are 0, 1, 2, 3. The number of possible samples of 3 bulbs from 12 is $\dbinom{12}{3} = \dfrac{12 \times 11 \times 10}{3 \times 2 \times 1} = 220$.

Further, these are all equally likely. The probabilities of the various values of X are:

$$P[X = 0] = \frac{\dbinom{8}{3}}{\dbinom{12}{3}} = \frac{56}{220}$$

$$P[X = 1] = \frac{\dbinom{4}{1} \times \dbinom{8}{2}}{\dbinom{12}{3}} = \frac{4 \times 28}{220} = \frac{112}{220}$$

$$P[X=2] = \frac{\binom{4}{2} \times \binom{8}{1}}{\binom{12}{3}} = \frac{6 \times 8}{220} = \frac{48}{220}$$

$$P[X=3] = \frac{\binom{4}{3}}{\binom{12}{3}} = \frac{4}{220}$$

For instance, $X = 1$ corresponds to the occurrence of 1 defective and 2 good bulbs in the sample. The number of possible samples that have 1 defective and 2 good bulbs is $\binom{4}{1} \times \binom{8}{2}$.

x	$f(x)$
0	0.255
1	0.509
2	0.218
3	0.018
Total	1.000

5.97 (a) & (b) We calculate $\mu = \sum x_i f(x_i)$ and calculate σ^2 using the alternative formula $\sigma^2 = \sum x_i^2 f(x_i) - \mu^2$:

x	$f(x)$	$xf(x)$	$x^2 f(x)$
2	0.1	0.2	0.4
3	0.3	0.9	2.7
4	0.3	1.2	4.8
5	0.2	1.0	5.0
6	0.1	0.6	3.6
		3.9	16.5

$E(X) = \mu = 3.9$

$\sigma^2 - 16.5 - (3.9)^2 = 1.29$, and $sd(X) = \sigma = \sqrt{1.29} = 1.14$

(c) The probability histogram of X with $\mu = E(X)$ located is given below.

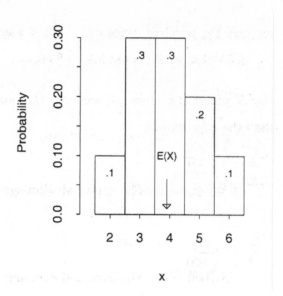

5.99 (a) The probability that the student will get either of the two winning tickets is
$$\frac{2}{1000} = 0.002.$$

(b) Consider X = dollar amount of the student's winnings. The random variable X can have the values 0 or 200 with probabilities 0.998 and 0.002, respectively.

x	$f(x)$	$xf(x)$
0	0.998	0
200	0.002	0.4
		0.4 $= E(X)$

Considering now the purchase price of $1, the student's expected gain
$= \$0.40 - \$1 = -\$0.60$, that is, expected loss $= \$0.60$.

5.101 (a) $P[X < 3] = f(0) + f(1) + f(2) = 0.05 + 0.1 + 0.15 = 0.30.$

(b)

x	$f(x)$	$xf(x)$	$x^2 f(x)$
0	0.05	0	0
1	0.10	0.10	0.10
2	0.15	0.30	0.60
3	0.35	1.05	3.15
4	0.20	0.80	3.20
5	0.15	0.75	3.75
Total		$3.00 = \mu$	10.80

$E(X) = 3.00$

$\sigma^2 = \sum x^2 f(x) - \mu^2 = 10.80 - (3.00)^2 = 1.80$

$\sigma = \sqrt{1.80} = 1.342$

5.103 (a) Let A, B, C denote the correct names. Listed below are the elementary outcomes, that is, the possible assignments of names, and the corresponding values of X = number of matches.

	Correct names			Value of
	A	B	C	X
	A	B	C	3
	A	C	B	1
Possible	B	A	C	1
assignments	B	C	A	0
	C	A	B	0
	C	B	A	1

(b) & (c) All 6 elementary outcomes are equally likely, Therefore,

$$P[X = 0] = \frac{2}{6}, \quad P[X = 1] = \frac{3}{6}, \quad P[X = 3] = \frac{1}{6}.$$

The probability distribution is presented below along with the calculation of its expectation.

x	$f(x)$	$xf(x)$
0	2/6	0
1	3/6	3/6
3	1/6	3/6
Total		$1 = E(X)$

5.105 (a) & (b) The possible values of X are:

$X = -15$ if he loses all three times

$X = 5 - 5 - 5 = -5$ if he loses twice and wins once

$X = 5 + 5 - 5 = 5$ if he loses once and wins twice

$X = 5 + 5 + 5 = 15$ if he wins all three times.

Using the symbol W for win and L for loss at each play we list the elementary outcomes. In assigning probabilities, note that at each play $P(W) = \dfrac{18}{38}$ and

$P(L) = \dfrac{20}{38}$ because 18 out of 38 slots are red, and 20 are not red. Also the outcomes at different plays are independent.

$$\underline{X = -15} \quad \underline{X = -5} \quad \underline{X = 5} \quad \underline{X = 15}$$

$$LLL \qquad WLL \qquad WWL \quad WWW$$
$$LWL \qquad WLW$$
$$LLW \qquad LWW$$

$P(LLL) = \left(\dfrac{20}{38}\right)^3 = 0.1458$ so $P[X = -15] = 0.1458$

$P(WLL) = \dfrac{18}{38} \times \dfrac{20}{38} \times \dfrac{20}{38}$, and the same result holds for LWL and LLW.

Therefore, $P[X = -5] = 3 \times \dfrac{18}{38} \times \left(\dfrac{20}{38}\right)^2 = 0.3936$

Similarly, $P[X = 5] = 3 \times \left(\dfrac{18}{38}\right)^2 \times \dfrac{20}{38} = 0.3543$, $P[X = 15] = \left(\dfrac{18}{38}\right)^3 = 0.1063$.

Probability distribution of X and calculation of expectation

x	$f(x)$	$xf(x)$
−15	0.1458	−2.187
−5	0.3936	−1.968
5	0.3543	1.7715
15	0.1063	1.5945
Total	1.0000	−0.789

$E(X) = -\$0.79$

(c) No. At any play, betting on red or black are probabilistically equivalent because in either case $P(W) = \dfrac{18}{38}$. Also, different plays are independent.

5.107 In solving Exercise 5.11, we already obtained the probability distribution of $(X_1 + X_2)$, the total number of dots resulting from two tosses of a fair die. Here we are concerned with the new random variable $\overline{X} = \dfrac{X_1 + X_2}{2}$. The distinct values of $X_1 + X_2$ lead to the distinct values of $\overline{X}$:

$$X_1 + X_2 \text{ values}: \quad 2 \quad 3 \quad 4 \quad 5 \quad \ldots \quad 11 \quad 12$$

$$\overline{X} \text{ values}: \quad 1 \quad 1.5 \quad 2 \quad 2.5 \quad \ldots \quad 5.5 \quad 6$$

The probability distribution of $\overline{X}$ is

$\overline{x}$	$f(\overline{x})$
1	1/36
1.5	2/36
2	3/36
2.5	4/36
3	5/36
3.5	6/36
4	5/36
4.5	4/36
5	3/36
5.5	2/36
6	1/36
Total	1

5.109 (a)

x	$f(x)$	$F(x)$
1	0.07	0.07
2	0.12	0.19
3	0.25	0.44
4	0.28	0.72
5	0.18	0.90
6	0.10	1.00

5.111

x	$f(x)$	$xf(x)$	$x^2 f(x)$
2	2/15	4/15	8/15
3	4/15	12/15	36/15
4	6/15	24/15	96/15
5	3/15	15/15	75/15
Total	1	55/15	215/15

$$\mu = \frac{55}{15} = 3.667$$

$$\sigma^2 = \frac{215}{15} - \left(\frac{55}{15}\right)^2 = 0.889$$

$$\sigma = \sqrt{0.899} = 0.943$$

5.113 (a) Although there are two possible outcomes for each trial, the model of Bernoulli trials is not appropriate. The independence assumption is questionable.

(b) Bernoulli model is plausible.

(c) Not Bernoulli trials because time has more than two possible outcomes.

(d) There are two possible outcomes but the condition of independence does not hold. Clear or cloudy condition often lasts over several days.

(e) Bernoulli model is plausible.

5.115 Denoting male by M and female by F, we calculate

$$P(FFM) = 0.5 \times 0.5 \times 0.5 = 0.125.$$

5.117 Let $S =$ hitting a red light. We have $P(S) = 0.7$ and $P(F) = 0.3$.

(a) $P(SS) = 0.7 \times 0.7 = 0.49$

(b) $P(SSS) = (0.7)^3 = 0.343$

(c) $P(SSF) + P(SFS) + P(FSS) = 3 \times (0.7)^2 \times 0.3 = 0.441.$

5.119 $P(FF) = q^2 = \dfrac{4}{49}$ so $q = \sqrt{\dfrac{4}{49}} = \dfrac{2}{7}$ and $p = 1 - q = \dfrac{5}{7}$

$$P(SSF) = p^2 q = \left(\frac{5}{7}\right)^2 \frac{2}{7} = \frac{50}{343} = 0.146.$$

5.121 (a) X has the binomial distribution with $n = 6$ and $p = 0.4$.

(b) Using the binomial table, we find

$$P[X \le 3] = 0.821, \; P[X = 0] = 0.047$$

$$E(X) = np = 6 \times 0.4 = 2.4 \text{ persons}$$

5.123 Let X = number of victims under 16 out of 14 moped accident victims. The distribution of X is binomial with $n = 14$ and $p = 0.33$.

(a) Mean $= np = 14 \times 0.33 = 4.62$

(b) $sd = \sqrt{npq} = \sqrt{14 \times 0.33 \times 0.67} = \sqrt{3.095} = 1.759$

(c) P[first victim is under 16 and second victim is at least 16]
  $= pq = 0.33 \times 0.67 = 0.221$.

5.125 Employing the binomial model with $n = 20$ and $p = 0.7$, we find $P[X \leq 10] = 0.048$. The probability of the observed result 10, or a more extreme result, is so small that we would doubt the claim that $p = 0.7$. Not as many students as claimed support the paper's view.

5.127 We use the binomial table for $n = 14$, $p = 0.4$.

(a) $P[3 \leq X \leq 9] = P[X \leq 9] - P[X \leq 2] = 0.982 - 0.040 = 0.942$

(b) $P[3 < X \leq 9] = P[X \leq 9] - P[X \leq 3] = 0.982 - 0.124 = 0.858$

(c) $P[3 < X < 9] = P[4 \leq X \leq 8] = P[X \leq 8] - P[X \leq 3] = 0.942 - 0.124 = 0.818$

(d) $E(X) = 14 \times 0.4 = 5.6$

(e) $sd(X) = \sqrt{14 \times 0.4 \times 0.6} = 1.833$.

5.129 P[1 success in n trials] $= \binom{n}{1} pq^{n-1} = npq^{n-1}$

P[0 successes in n trials] $= q^n$

One success is more probable than 0 successes if $npq^{n-1} > q^n$ or

$$n > \frac{q^n}{pq^{n-1}} = \frac{q}{p} = \frac{0.85}{0.15} = 5.67$$

The smallest n (integer) that satisfies $n > 5.67$ is 6.

5.131 (a) $n = 12$

p	0.1	0.2	0.3	0.4	0.5
$P[X \leq 3]$	0.974	0.795	0.493	0.225	0.073

(b) $n = 18$

p	0.1	0.2	0.3	0.4	0.5
$P[X \leq 3]$	0.902	0.501	0.165	0.033	0.004

5.133 Let X = number of persons (out of 20) who feel the system is adequate. Since the city population is large, the binomial distribution is appropriate for X. We have $n = 20$ and $p = 0.3$ so

$$P[X \geq 10] = 1 - P[X \leq 9] = 1 - 0.952 = 0.048$$
$$P[X = 10] = P[X \leq 10] - P[X \leq 9] = 0.983 - 0.952 = 0.031.$$

5.135 (a) Let X = amount of bonus money received. Then X can take on two values: 0, 500. Let Y = the number of months out of 3 months in which she recruits a new client. The distribution of Y is binomial with $n = 3$, $p = 0.4$. Then

$$P[X = 0] = P[Y \leq 2] = 0.936 \quad P[X = 500] = P[Y = 3] = 0.064$$

The probability distribution of X is given below.

x	$f(x)$
0	0.936
600	0.064

(b) Using the table in part (a)

$$E(X) = \sum xf(x) = 0 \times 0.936 + 600 \times 0.064 = 38.4 \text{ dollars}$$

5.137 (a) $P[X = 0] = f(0) = e^{-3} \dfrac{(3)^0}{0!} = e^{-3} = 0.05$

(b) $P[X = 1] = f(1) = e^{-3} \dfrac{(3)^1}{1!} = e^{-3} \times 3 = 0.15.$

Chapter 6

THE NORMAL DISTRIBUTION

6.1 (a) The function is non-negative and the area of the rectangle is $0.5 \times 2 = 1$. It is a probability density function.

(b) Since $f(x)$ takes negative values over the interval from 1 to 2, it is not a probability density function.

(c) The function is non-negative and the area of the triangle is $\frac{1}{2} \times base \times height = \frac{1}{2} \times 2 \times 1 = 1$. It is a probability density function.

(d) The function is non-negative, but the area of the rectangle is $1 \times 2 = 2$ so it is not a probability density function.

6.3 From the picture of $f(x)$ we see that the interval 1.5 to 2 has a larger area under $f(x)$ than the interval 0 to 0.5. Consequently, $P[1.5 < X < 2]$ is the larger of the two probabilities.

6.5 For an arbitrary point x in the interval 0 to 2, we find that

$$P[0 < X < x] = \text{Area of the triangle over the base } (0,x) \ = \frac{1}{2} \cdot x \cdot \frac{x}{2} = \frac{x^2}{4}$$

The median is the value of x for which the cumulative probability $x^2/4 = 0.5$ so $x = \sqrt{2} = 1.414$. Similarly, $x = \sqrt{1} = 1$ is Q_1 and $x = \sqrt{3} = 1.73$ is Q_3.

6.7 (a) The median is the time such that there is an equal probability of being earlier or of being later. Consequently, the median is later than 1:20 p.m.

(b) No. The mean could be larger or smaller than the median depending on the distribution. The mean is often larger than the median when the distribution has a long tail to the right.

6.9 The standardized variable is $Z = \dfrac{X - \mu}{\sigma}$. So,

(a) $Z = (X - 15)/4$

(b) $Z = (X - 61)/9$

(c) $Z = (X - 161)/\sqrt{25} = (X - 161)/5$

6.11 (a) $Z = (X - 8)/2$

(b) $Z = (X - 350)/8$

(c) $Z = (X - 888)/\sqrt{81} = (X - 888)/9$

6.13 We use Appendix B Table 3 which gives the area under the standard normal curve to the left of a z-value.

(a) $P[Z < 0.83] = 0.7967$

(b) $P[Z < 1.03] = 0.8485$

(c) $P[Z < -1.03] = 0.1515$

(d) $P[Z < -1.35] = 0.0885$

6.15 The area to the right of a z-value $= 1 -$ Area to the left of the z-value.

(a) $P[Z > 0.83] = 1 - P[Z < 0.83] = 1 - 0.7967 = 0.2033$

(b) $1 - P[Z < 2.83] = 1 - 0.9977 = 0.0023$

(c) $1 - P[Z < -1.23] = 1 - 0.1093 = 0.8907$

(d) $z = 1.635$ is the mid-point between $z = 1.63$ and $z = 1.64$. From the normal table, we find

$$
\begin{array}{ll}
P[Z < 1.64] & = 0.9495 \\
P[Z < 1.63] & = 0.9484 \\
\hline
\text{difference} & = 0.0011
\end{array}
$$

The mid-point between 0.9495 and 0.9484 is

$$0.9484 + \frac{1}{2}(0.0011) = 0.9490$$

so $P[Z \leq 1.635] = 0.9490$ and $P[Z > 1.635] = 1 - 0.9490 = 0.0510$.

6.17 (a) $P[-0.44 < Z < 0.44] = P[Z < 0.44] - P[Z < -0.44] = 0.6700 - 0.3300 = 0.3400$.

(b) $P[-1.33 < Z < 1.33] = P[Z < 1.33] - P[Z < -1.33] = 0.9082 - 0.0918 = 0.8164$.

(c) $P[0.40 < Z < 2.03] = P[Z < 2.03] - P[Z < 0.40] = 0.9788 - 0.6554 = 0.3234$.

(d) We find that

$$P[Z < 1.41] \quad = \quad 0.9207$$
$$P[Z < 1.40] \quad = \quad 0.9192$$
$$\overline{\text{difference} \quad = \quad 0.0015}$$

Therefore,

$$P[Z < 1.405] = 0.9192 + \frac{1}{2}(0.0015) = 0.9200 \text{ (rounded)}$$

Also,

$$P[Z < 2.31] \quad = \quad 0.9896$$
$$P[Z < 2.30] \quad = \quad 0.9893$$
$$\overline{\text{difference} \quad = \quad 0.0003}$$

so

$$P[Z < 2.306] = 0.9893 + 0.6 \times (0.0003) = 0.9895 \text{ (rounded)}$$

Finally,

$$P[1.405 < Z < 2.306] = 0.9895 - 0.9200 = 0.0695$$

6.19 (a) We are to find the z-value for which the area to the left is 0.20. From the normal table, we find

$$P[Z < -0.84] \quad = \quad 0.2005$$
$$P[Z < -0.85] \quad = \quad 0.1977$$
$$\overline{\text{difference} \quad = \quad 0.0028}$$

Since $0.2005 - 0.20 = 0.0005$, the required z-value is

$$-0.84 - (0.01) \times \frac{0.0005}{0.0028} = -0.842$$

(b) We are to find the z-value for which the area to the left is $1 - 0.125 = 0.875$ so $z = 1.15$.

(c) By the symmetry of the normal curve, we have

$$2P[Z < -z] = 1 - P[-z < Z < z] = 1 - 0.668 - 0.332$$

or

$$P[Z < -z] = 0.166, \text{ so that } z = 0.97.$$

(d) We are to find the z-value for which

$$0.888 = P[z < Z < 2.0] = P[Z < 2.0] - P[Z < z] = 0.9772 - P[Z < z]$$

or

$$P[Z < z] = 0.0892.$$

We find that

$$P[Z < -1.34] = 0.0901$$
$$P[Z < -1.35] = 0.0885$$
$$\overline{\text{difference} \quad = 0.0016}$$

The required z-value is

$$-1.34 - (0.01) \times \frac{0.0009}{0.0016} = -1.34 - 0.006 = -1.346$$

6.21 (a) $P[Z < -0.93] = 0.1762$ so the z-value is -0.93.

(b) We are to find the z-value for which the area to the left is $1 - 0.10 = 0.90$. From the normal table, we find that

$$P[Z < 1.29] = 0.9015$$
$$P[Z < 1.28] = 0.8997$$
$$\overline{\text{difference} \quad = 0.0018}$$

since 0.8997 is nearly 0.90 we could take $z = 1.28$. More accurately, the required z-value is

$$1.28 + (0.01) \times \frac{0.0003}{0.0018} = 1.28 + 0.0017 = 1.2817 \text{ or } 1.282 \text{ (rounded)}$$

(c) Since $2P[Z < -z] = 1 - 0.954 = 0.046$, we require that

$$P[Z < -z] = \frac{1}{2}(0.046) = 0.023 . \text{ From the normal table, we see that}$$

$$P[Z < -1.99] = 0.0233$$
$$P[Z < -2.00] = 0.0228$$
$$\overline{\text{difference} \quad = 0.0005}$$

so

$$-1.99 - (0.01) \times \frac{0.0003}{0.0005} = -1.996$$

is the required z-value.

(d) We require that

$$P[Z < z] = 0.50 + P[Z < -0.6] = 0.50 + 0.2743 = 0.7743$$

Scanning the normal table, we find

$$P[Z < 0.76] = 0.7764$$
$$P[Z < 0.75] = 0.7734$$
$$\overline{\text{difference} \quad = 0.0030}$$

The probability 0.7743 is about a third of the way, so we could use 0.753. More accurately, the z-value is

$$0.75 + (0.01) \times \frac{0.0009}{0.0030} = 0.753$$

6.23 (a) $P[Z < 0.33] = 0.6293$

(b) We find $P[Z < -0.44] = 0.3300$ Therefore, the 33th percentile is -0.44.

(c) $P[Z < 0.70] = 0.7580$

(d) We find that

$$P[Z < 0.53] \quad = \ 0.7019$$
$$\underline{P[Z < 0.52] \quad = \ 0.6985}$$
$$\text{difference} \quad = \ 0.0034$$

Therefore, the 70th percentile is

$$0.52 + (0.01) \times \frac{0.0015}{0.0034} = 0.524$$

6.25 The standardized variable is $Z = \dfrac{X - 60}{4}$.

(a) For $x = 55$, we have $z = (55 - 60)/4 = -1.25$ so
$$P[X < 55] = P[Z < -1.25] = 0.1056$$

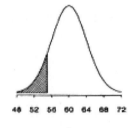 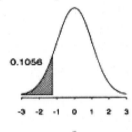

(b) $P[X \leq 67] = P[Z \leq 1.75] = 0.9599$

(c) $P[X > 66] = P[Z > 1.50] = 1 - P[Z \leq 1.50] = 1 - 0.9332 = 0.0668$

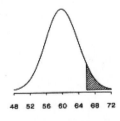 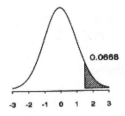

(d) $P[X > 51] = P[Z > -2.25] = 1 - 0.0122 = 0.9878$

(e) $P[53 \leq X \leq 69] = P[\dfrac{53 - 60}{4} \leq Z \leq \dfrac{69 - 60}{4}] = P[-1.75 \leq Z \leq 2.25]$

$$= P[Z < 2.25] - P[Z < -1.75] = 0.9878 - 0.0401 = 0.9477$$

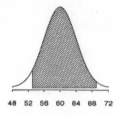

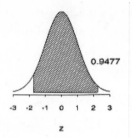

(f) $P[61 < X < 64] = P[0.25 < Z < 1.00] = 0.8413 - 0.5987 = 0.2426$

6.27 The standardized variable is $Z = \dfrac{X - 130}{5}$.

    (a) From the normal table we find $P[Z < 1.96] = 0.975$. Therefore, $\dfrac{b - 130}{5} = 1.96$

    so that $b = 130 + 5(1.96) = 139.8$.

    (b) Here, $P[X < b] = 1 - 0.025 = 0.975$. As calculated in part (a), we have
    $b = 139.8$

    (c) Since $P[Z < -0.51] = 0.305$, we must have
$$\frac{b - 130}{5} = -0.51 \text{ or } b = 120 + 5(-0.51) = 117.45.$$

6.29 Let X denote the score of a randomly selected student. We have $\mu = 500$ and
    $\sigma = 100$ so the standardized variable is $Z = \dfrac{X - 500}{100}$.

    (a) $P[X > 650] = P\left[Z > \dfrac{650 - 500}{100}\right] = P[Z > 1.5] = 1 - 0.9332 = 0.0668$

    (b) $P[X < 250] = P[Z < -2.5] = 0.0062$

    (c) $x = 325$ gives $z = \dfrac{325 - 500}{100} = -1.75$

        $x = 675$ gives $z = \dfrac{675 - 500}{100} = 1.75$

    Thus, we have
    $P[325 < X < 675] = P[-1.75 < Z < 1.75] = 0.9599 - 0.0401 = 0.9198$.
    These are illustrated on the following page.

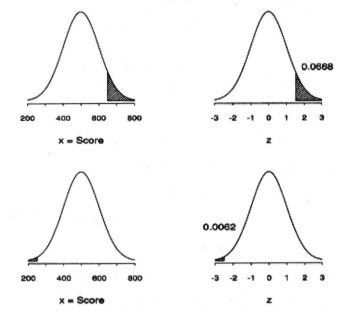

6.31 The z-value for 32.5 inches is $(32.5-34.5)/1.3 = -1.538$ and that for 36.5 inches is $(36.5-34.5)/1.3 = 1.538$ so
$$P[32.5 < X < 36.5] = P[-1.538 < Z < 1.538] = 0.9378 - 0.0623 = 0.8755$$

6.33 The z-value for 4 ounces is $(4-5)/1.2 = -0.833$, so that
$$P[X < 4] = P[Z < -0.833] = 0.2024$$

6.35 The standardized variable is $Z = \dfrac{X - 64.1}{3.4}$.

(a) The z-value for 70 inches is $(70-64.1)/3.4 = 1.735$. Consequently, by interpolating in the table, we find that
$$P[X > 70] = P[Z > 1.735] = 1 - P[Z < 1.735] = 1 - 0.9587 = 0.0413.$$

(b) The z-value for 60 inches is $(60-64.1)/3.4 = -1.206$. Consequently (again using interpolation in the table),
$$P[X \le 60] = P[Z \le -1.206] = 0.1139.$$

6.37 The arrival time X is distributed as $N(17,3)$.

(a) (i) $\quad P[X > 22] = P[Z > \dfrac{22-17}{3}] = P[Z > 1.67] = 1 - 0.9525 = .0475$

(ii) The z-values corresponding to $x = 13$ and $x = 21$ are $\dfrac{13-17}{3} = -1.33$ and

$\dfrac{21-17}{3}=1.33$, respectively. Thus, we have

$$P[13 < X < 21] = P[-1.33 < Z < 1.33] = 0.9082 - 0.0918 = 0.8164$$

(iii) Since $\dfrac{15.5-17}{3}=-0.5$, $\dfrac{18.5-17}{3}=0.5$, we have

$$P[15.5 < X < 18.5] = P[-0.5 < Z < 0.5] = 0.6915 - 0.3085 = 0.3830$$

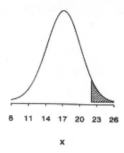
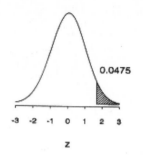

(b) The probability density curve of a normal distribution peaks at the mean.
Therefore, the 1-minute interval that has the highest probability is one that is
centered at $\mu = 17$ and has length 1, that is, the interval 16.5 to 17.5.

6.39 (a) We use the binomial table (Appendix B Table 2) for $n = 25$ and $p = 0.6$.

(i) $P[X = 17] = P[X \le 17] - P[X \le 16] = 0.846 - 0.726 = 0.120$

(ii) $P[11 \le X \le 18] = P[X \le 18] - P[X \le 10] = 0.926 - 0.034 = 0.892$

(iii) $P[11 < X < 18] = P[X \le 17] - P[X \le 11] = 0.846 - 0.078 = 0.768$

(b) With $n = 25$ and $p = 0.6$, we calculate

$$np = 15 \text{ and } \sqrt{npq} = \sqrt{25 \times 0.6 \times 0.4} = 2.45 .$$

We consider X to be normally distributed with mean $= 15$ and $sd = 2.45$, so

$$Z = \dfrac{X - 15}{2.45} \approx N(0,1)$$

(i)   Using the continuity correction, we calculate the normal probability
assigned to the interval 16.5 to 17.5.

$$P[16.5 < X < 17.5] = P[\dfrac{16.5-15}{2.45} < Z < \dfrac{17.5-15}{2.45}]$$
$$= P[0.612 < Z < 1.020]$$
$$= 0.8461 - 0.7298 = 0.1163.$$

(ii)   With continuity correction, the relevant interval is 10.5 to 18.5.

$$P[10.5 < X < 18.5] = P[\frac{10.5-15}{2.45} < Z < \frac{18.5-15}{2.45}]$$
$$= P[-1.837 < Z < 1.429]$$
$$= 0.9235 - 0.0331 = 0.8904.$$

(iii) Here the end points 11 and 18 are not included. The continuity correction leads to the interval 11.5 to 17.5.

$$P[11.5 < X < 17.5] = P[-1.429 < Z < 1.020]$$
$$= 0.8461 - 0.0765 = 0.7696.$$

6.41 To use the normal approximation, we calculate

$$\text{mean} = np = 300 \times 0.25 = 75$$
$$sd = \sqrt{npq} = \sqrt{300 \times 0.25 \times 0.75} = 7.50$$

so $Z = \dfrac{X-75}{7.50}$ is approximately $N(0, 1)$.

(a) We calculate the normal probability of the interval 79.5 to 80.5.
$$P[X = 80] \approx P[\frac{79.5-75}{7.50} < Z < \frac{80.5-75}{7.50}]$$
$$= P[0.60 < Z < 0.733] = 0.7683 - 0.7257 = 0.0426$$

(b) Using the continuity correction,
$$P[X \le 65] \approx P[Z < \frac{65.5-75}{7.50}] = P[Z < -1.267] = 0.1025$$

(c) $P[68 \le X \le 89] \approx P[\frac{67.5-75}{7.50} < Z < \frac{89.5-75}{7.50}]$
$$= P[-1.0 < Z < 1.933]$$
$$= 0.9734 - 0.1587 = 0.8147$$

6.43 (a) Normal approximation is appropriate because n is large and p is not too close to 0 or 1.
(b) Not appropriate because p is too small, $np = 3$.
(c) Not appropriate because p is too close to 1, $n(1-p) = 2.4$.
(d) Normal approximation is appropriate because n is large and p is not too close to 0 or 1.

6.45 The standardized scale is $z = \dfrac{x - np}{\sqrt{np(1-p)}}$. We have the following:

$$n = 5, \quad p = 0.4: \quad z = (x-2)/1.095$$
$$n = 12, \quad p = 0.4: \quad z = (x-4.8)/1.697$$
$$n = 25, \quad p = 0.4: \quad z = (x-10)/2.449.$$

The probability histograms of binomial distribution for $p = 0.4$, $n = 5, 12$, and 25, and the corresponding z-scores are given below.

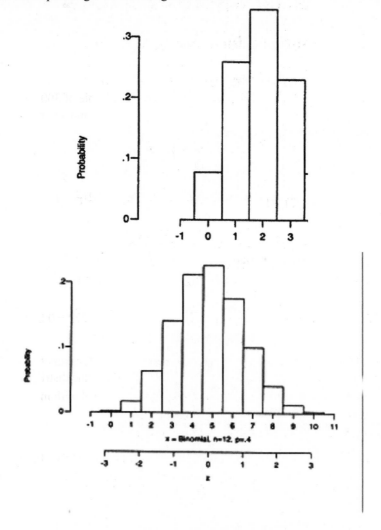

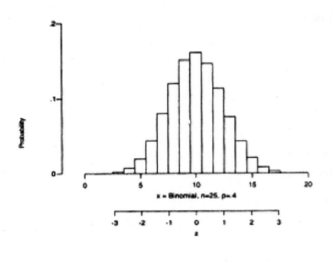

6.47 Let X = number of unemployed persons in a random sample of 300. Then, the distribution of X is binomial with $n = 300$ and $p = 0.079$, and a normal approximation is appropriate. Here

$$np = 300 \times 0.079 = 23.7$$

$$\sqrt{npq} = \sqrt{300 \times 0.079 \times 0.921} = 4.672$$

$$Z = \frac{X - 23.7}{4.672}$$

(a) $P[X < 18] \approx P[Z < \frac{17.5 - 23.7}{4.672}] = P[Z < -1.327] = 0.0923$

(b) $P[X > 30] \approx P[Z > \frac{30.5 - 23.7}{4.672}] = P[Z > 1.455] = 1 - 0.9272 = 0.0728$.

6.49 We assume the customers make purchase independently of one another. The binomial distribution applies. Since $n = 80$ and $p = 0.20$, the distribution of X = number who make a purchase, is approximately normal with mean $= 80 \times 0.20 = 16.0$ and $sd = \sqrt{80 \times 0.20 \times 0.80} = 3.578$.

Using continuity correction, we approximate

$$P[x > 20] = P[Z > \frac{20.5 - 16.0}{3.578}] = 1 - P[Z > 1.258] = 1 - 0.8958 = 0.1042$$

6.51 Let Y be next weeks expenditure so

$$p = P[Y > 880] = P[Z > \frac{880 - 850}{40}] = P[Z > 0.75] = 1 - 0.7734 = 0.2266$$

Let X be the number of weeks, out of $n = 52$, where, the expenses exceed 580 dollars. Then X has the binomial distribution with $n = 52$ and $p = 0.2266$. To use the normal approximation, we calculate

$$\mu = np = 52 \times 0.2266 = 11.783 \text{ and } \sigma = \sqrt{npq} = \sqrt{52 \times 0.2266 \times 0.7734} = 3.019.$$

Using continuity correction,

$$P[10 \le X \le 16] \approx P[\frac{9.5 - 11.783}{3.019} \le Z \le \frac{16.5 - 11.783}{3.019}]$$
$$= P[-0.756 < Z < 1.562] = 0.9409 - 0.2248 = 0.7161.$$

6.53 We assume the residents make their decision to move independently of one another. The binomial distribution applies. Since $n = 100$ and $p = 0.33$, the distribution of X = number who move, is approximately normal with mean $= 100 \times 0.33 = 33$ and $sd = \sqrt{100 \times 0.33 \times 0.67} = 4.702$

Using continuity correction, we approximate

$$P[X \ge 39] \approx P[Z > \frac{38.5 - 33}{4.702}] = 1 - P[Z < 1.170] = 1 - 0.8790 = 0.1210$$

6.55 The probability density curve is a rectangle with height $= 1$.

(a) The point $x = 0.5$ divides the distribution into halves so median $= 0.5$.
(b) The points $x = 0.25, 0.5$ and 0.75 divide the distribution into quarters. Therefore, first quartile $= 0.25$, second quartile $= 0.5$, and third quartile $= 0.75$.

6.57 We could make a histogram using the first 100 heights and then another based on the first 500 heights. The histogram at the front of the chapter is based on 1,456 observations, a large number. From these original data of heights, one can construct a histogram with considerably smaller class intervals than given in the chapter. If this process is repeated with an even much larger sample, the jumps between consecutive rectangles will then dampen out, and the top of the histogram will approximate the shape of a smooth curve. The density curve would be non-negative and the area under the curve is one.

6.59 (a) Scanning the probabilities given in the normal table, we find that $P[Z < -1.38] = 0.0838$. Therefore, $z = -1.38$.
(b) In the table we look for probabilities close to .047, and find $P[Z < -1.68] = 0.0465$ and $P[Z < -1.67] = 0.0475$ Interpolation gives $P[Z < -1.675] = 0.047$, so that $z = -1.675$.

(c) Area to the left of z is $1 - 0.2611 = 0.7389$. From the table we get
$$P[Z < 0.64] = 0.7389, \text{ so that } z = 0.64.$$

(d) Area to the left of z is $1 - 0.12 = 0.88$. The table gives
$$P[Z < 1.17] = 0.8790 \text{ and } P[Z < 1.18] = 0.8810.$$
Since 0.88 is halfway between these results, we interpolate $z = 1.175$.

6.61 (a) $P[Z > 0.62] = 1 - P[Z < 0.62] = 1 - 0.7324 = 0.2676$

(b) $P[-1.40 < Z < 1.40] = P[Z < 1.40] - P[Z < -1.40] = 0.9192 - 0.0808 = 0.8384$

(c) $P\big[|Z| > 3\big] = P\big[Z > 3\big] + P\big[Z < -3\big]$
$$= 2P\big[Z < -3\big] \qquad \text{(by symmetry)}$$
$$= 2 \times 0.0013 = 0.0026$$

(d) $P\big[|Z| < 2\big] = P\big[-2 < Z < 2\big] = P\big[Z < 2\big] - P\big[Z < -2\big] = 0.9772 - 0.0228 = 0.9544$

Alternatively, we can calculate
$$P\big[|Z| > 2\big] = 2P\big[Z < -2\big] \qquad \text{(as in part (b))}$$
$$= 2 \times 0.0228 = 0.0456$$

Hence, $P\big[|Z| < 2\big] = 1 - 0.0456 = 0.9544$.

6.63 The standard normal variable is $Z = \dfrac{X - \mu}{\sigma} = \dfrac{X - 100}{8}$.

(a) $P[X < 107] < P[\dfrac{107 - 100}{8}] = P[Z < 0.875] = 0.8092$

(b) $P[X < 97] = P[Z < -0.375] = 0.3539$

(c) $P[X > 110] = P[Z > 1.25] = 1 - 0.8944 = 0.1056$

(d) $P[X > 90] = P[Z > -1.25] = 1 - 0.1056 = 0.8944$

(e) $P[95 < Z < 106] = P[-0.625 < Z < 0.75] = 0.7734 - 0.2660 = 0.5074$

(f) $P[103 < X < 114] = P[0.375 < Z < 1.75] = 0.9599 - 0.6461 = 0.3138$

(g) $P[88 < X < 100] = P[-1.5 < Z < 0] = 0.5 - 0.0668 = 0.4332$

(h) $P[60 < X < 108] = P[-5 < Z < 1] = 0.8413 - 0 = 0.8413$

6.65 The standardized variable is $Z = \dfrac{X - 497}{120}$.

(a) $P[X > 600] = P[Z > \dfrac{600 - 497}{120}] = P[Z > 0.858] = 1 - 0.8045 = 0.1955$

(b) We first find the 90th percentile of the standard normal distribution and then convert it to the x scale. Indeed, observe that $P[Z < 1.28] = 0.8997 \approx 0.90$
The standardized score $z = 1.28$ corresponds to $x = 497 + 120(1.28) = 651$

(c) $P[X < 400] = P[Z < -0.808] = 0.2096$.

6.67 The strength X is distributed as $N(100,8)$. The bonding will fail if $X < 98$. Its probability is

$$P[X < 98] = P[Z < \frac{98-100}{8}] = P[Z < -0.25] = 0.4013.$$

6.69 X = number of days of trouble-free operation is normally distributed with $\mu = 530$ and $\sigma = 100$, so the standardized variable is $Z = \frac{X-530}{100}$.

(a) Taking 365 days in a year, 2 years have 730 days.

$$P[X > 730] = P[Z > \frac{730-530}{100}] = P[Z > 2] = 1 - 0.9772 = 0.0228.$$

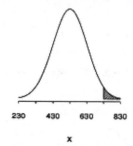 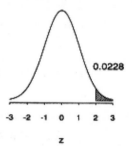

(b) Denoting d = number of days of warranty, we require that $P[X < d] = 0.1$. From the standard normal table, we find $P[Z < -1.28] = 0.10$ so

$$\frac{d-530}{100} = -1.28 \text{ or } d = 530 + 100(-1.28) = 402$$

The warranty period can be set to 402 (or about 400) days.

6.71 (a) Denote the volume in one bottle by X which is a random variable having a normal distribution with mean 302 and standard deviation 2 ml. We want to find $P[X < 299]$. Since $Z = \frac{X-302}{2}$ is standard normal, we have

$$P[X < 299] = P[Z < \frac{299-302}{2}] = P[Z < -1.5] = 0.0668$$

(b) We first determine that the z-value 1.645 satisfies $P[Z \le 1.645] = 0.9500$ so $P[Z > 1.645] = 0.05$. Consequently, $v = 302 + (2)1.645 = 305.3$ ounces.

6.73 (a) We use Appendix Table 2 to find the exact binomial probabilities.

(i) When $n = 25$, $p = 0.4$, $P[X \le 6] = 0.074$
(ii) When $n = 20$, $p = 0.7$,
$P[10 \le X \le 18] = P[X \le 18] - P[X \le 9] = 0.992 - 0.017 = 0.975$
(iii) When $n = 16$, $p = 0.5$,
$$P[X \ge 12] = 1 - P[X \le 11] = 1 - 0.962 = 0.038.$$

(b)   We take X to be approximately normally distributed with mean $= np$ and $sd = \sqrt{npq}$, and use the continuity correction. Also, we round the normal probabilities to three decimals.

(i) $np = 10, \sqrt{npq} = \sqrt{25 \times 0.4 \times 0.6} = 2.45$

$$P[X \le 6] = P[Z < \frac{6.5 - 10}{2.45}] = P[Z < -1.429] = 0.077$$

(ii) $np = 14, \sqrt{npq} = 2.05$

$$P[10 \le X \le 18] \approx P[\frac{9.5 - 14}{2.05} < Z < \frac{18.5 - 14}{2.05}]$$
$$= P[-2.195 < Z < 2.195] = 0.986 - 0.014 = 0.972$$

(iii) $np = 8, \sqrt{npq} = 2$

$$P[X \ge 12] = P[Z > \frac{11.5 - 8}{2}] = P[Z > 1.75] = 0.0401 .$$

6.75 Assuming $n = 400$ is a small fraction of the population, a binomial model is appropriate for $X =$ number of viewers of program A out of 400.

(a) The distribution of X is approximately normal with mean $= np = 400 \times 0.3 = 120$ and $sd = \sqrt{npq} = 9.165$.

$$P[X < 105] \approx P[Z < \frac{104.5 - 120}{9.165}] = P[Z < -1.69] = 0.0455$$

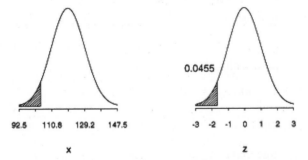

(b) Under the hypothesis that $p = 0.3$, the calculations in part (a) show that the occurrence of fewer than 105 viewers in a sample of 400 is unlikely (probability 0.0455). Such an observation should strongly support the suspicion that $p < 0.3$.

6.77 Out of 400 ticket holders, let $X =$ no. of persons who show up. The distribution of X is binomial with $n = 400$ and $p = 1 - 0.10 = 0.90$. It is approximately normal with mean $= np = 360$ and sd $= \sqrt{npq} = 6$.

(a) One or more reservation holders will not be accommodated if more than 370 passengers show up.
$$P[X > 370] \approx P[Z > \frac{370.5 - 360}{6}] = P[Z > 1.75] = 1 - 0.9599 = 0.0401$$

(b) $P[X < 350] \approx P[Z < \frac{349.5 - 360}{6}] = P[Z < -1.75] = 0.0401.$

6.79 Let X be the number of new words in the poem. We approximate the probability that a new word will not be on the list by $14,376/884,647 = 0.01625$. Then, we treat X as having a binomial distribution with $n = 429$ and $p = 0.01625$.

(a) Expected number of new words $=$ number of words $\times$ probability
$$= 429 \times 0.01625 = 6.97$$

(b) The standard deviation of X is $\sqrt{429 \times 0.01625 \times 0.98375} = 2.619$. Using the continuity correction, we approximate the binomial probability
$$P[X \geq 12] \approx P[Z \geq \frac{11.5 - 6.97}{2.619}] = P[Z \geq 1.73] = 1 - 0.9582 = 0.0418$$

(c) Using continuity correction, we approximate the binomial probability
$$P[X \leq 2] \approx P[z \leq \frac{2.5 - 6.97}{2.619}] = P[Z \leq -1.707] = 0.0439$$

(d) $P[2 < X < 12] \approx P[\frac{2.5 - 6.97}{2.619} \leq Z \leq \frac{11.5 - 6.97}{2.619}]$
$$= P[Z \leq \frac{11.5 - 6.97}{2.619}] - P[Z \leq \frac{2.5 - 6.97}{2.619}]$$
$$= P[Z \leq 1.73] - P[Z \leq -1.707]$$
$$= 0.9582 - 0.0439 = 0.9143$$

The observed number 9 is close to the expected number 6.97. The difference $9 - 6.97 = 2.03$ is not large according to our probability calculation which shows that ± 4 words from 6.97 is not unusual.

6.81 The normal scores plot for *volume* displays a curved pattern. The data are not normal.

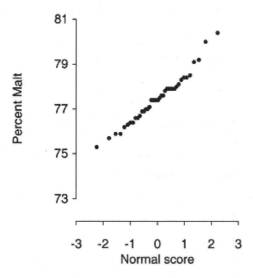

6.83 The data $\sqrt{(volume)}$ are not quite as normal looking as the fourth root, as illustrated in the following figure.

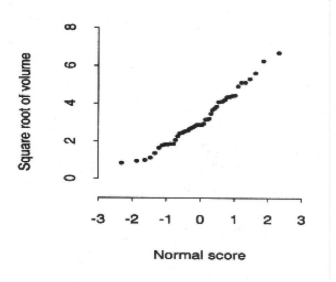

Chapter 7

VARIATION IN REPEATED SAMPLES –
SAMPLING DISTRIBUTIONS

7.1 (a)   Statistic   (b)   Statistic   (c)   Parameter
     (d)   Statistic   (e)   Parameter

7.3 (a) Students working full time would be more likely to take an evening course than students who do not work, and so they would be over-represented in the sample. Also, those who work the evening shift are left out of the survey.

(b) A large majority of persons, at all income levels, spend more during the holiday season. The total amount spent and the types of purchases are often atypical. The survey would likely be misleading.

7.5 (a) All possible samples (x_1, x_2) and the corresponding $\bar{x}$ values are:

(x_1, x_2)	(3,3)	(3,5)	(3,7)	(5,3)	(5,5)	(5,7)	(7,3)	(7,5)	(7,7)
$\bar{x} = \frac{x_1 + x_2}{2}$	3	4	5	4	5	6	5	6	7

(b) The 9 possible samples are equally likely, so each has a probability 1/9 of occurring. The sampling distribution of $\bar{X}$ is obtained by listing the distinct values of $\bar{X}$ along with the corresponding probabilities, as follows:

$\bar{x}$	Probability $f(\bar{x})$
3	1/9
4	2/9
5	3/9
6	2/9
7	1/9
Total	1

7.7   Not a random sample. The photographer would show his better pictures when trying to get a contract for wedding pictures.

7.9   (a) $X = 1$ if 1 or 2 dots show
          $X = 2$ if 3, 4 or 5 dots show
          $X = 4$ if 6 dots show

(b) Answers will vary. We obtained the following data from our experiment.

Roll 1 = 3 (so $X = 2$),   Roll 2 = 4 (so $X = 2$),   Roll 3 = 4 (so $X = 2$)

The median X value is therefore 2.

(c) Answers will vary. The data we collected for the 75 rolls of the die, grouped into 25 3-samples, are tabulated below, along with the corresponding values of X and the median of X in each case.

Rolls	Values of X	Median	Rolls	Values of X	Median
(3,4,4)	(2,2,2)	2	(2,2,2)	(1,1,1)	1
(2,1,6)	(1,1,4)	1	(5,3,2)	(2,2,1)	2
(6,1,1)	(4,1,1)	1	(1,1,2)	(1,1,1)	1
(3,3,4)	(2,2,2)	2	(6,4,6)	(4,2,4)	4
(6,1,6)	(4,1,4)	4	(5,5,2)	(2,2,1)	2
(5,3,2)	(2,2,1)	2	(1,1,6)	(1,1,4)	1
(2,2,6)	(1,1,4)	1	(3,3,5)	(2,2,2)	2
(3,1,3)	(2,1,2)	2	(4,1,2)	(2,1,1)	1
(3,4,1)	(2,2,1)	2	(5,1,1)	(2,1,1)	1
(6,2,2)	(4,1,1)	1	(3,2,3)	(2,1,2)	2
(3,5,5)	(2,2,2)	2	(3,1,5)	(2,1,2)	2
(3,6,4)	(2,4,2)	2	(2,4,4)	(1,2,2)	2
(4,4,3)	(2,2,2)	2			

The relative frequency distribution is:

X	Frequency	Relative Frequency	Population Relative Frequency
1	29	29/75 = 0.38667	0.33333
2	36	36/75 = 0.48000	0.50000
4	10	10/75 = 0.13333	0.16667
Total	75	1.0	1.0

The *sample* relative frequency distribution and the *population* relative frequency distribution should be close since the sample size of 75 is relatively large. In fact, as the sample size increases, the *sample* relative frequency distribution should better approximate the *population* relative frequency distribution.

(d) The sample median values are also listed in the table in part (c). The relative frequency distribution for the median values is as follows:

X_{median}	Frequency	Relative Frequency
1	9	9/25 = 0.36
2	14	14/25 = 0.56
4	2	2/25 = 0.08
Total	25	1.0

This distribution does approximate the sampling distribution since the 25 medians corresponding to the 3-samples of rolls are themselves values of X. As such, this is equivalent to rolling a die 25 times, assigning the appropriate value of X, and then forming a sampling distribution of X based on a sample of size 25. As mentioned above, as the sample size increases, such a *sample* relative frequency distribution should better approximate the *population* relative frequency distribution.

7.11 We have $\mu = 79$ and $\sigma = 9$.

(a) For $n = 4$, $E(\overline{X}) = \mu = 79$ and $\mathrm{sd}(\overline{X}) = \frac{\sigma}{\sqrt{n}} = \frac{9}{\sqrt{4}} = 4.5$.

(b) For $n = 25$, $E(\overline{X}) = \mu = 79$ and $\mathrm{sd}(\overline{X}) = \frac{\sigma}{\sqrt{n}} = \frac{9}{\sqrt{25}} = 1.8$.

7.13 The population standard deviation is $\sigma = 20$, so $\mathrm{sd}(\overline{X}) = \frac{\sigma}{\sqrt{n}} = \frac{20}{\sqrt{n}}$.

(a) For $n = 25$, $\mathrm{sd}(\overline{X}) = \frac{\sigma}{\sqrt{n}} = \frac{20}{\sqrt{25}} = 4.0$

(b) For $n = 100$, $\mathrm{sd}(\overline{X}) = \frac{\sigma}{\sqrt{n}} = \frac{20}{\sqrt{100}} = 2.0$

(c) For $n = 400$, $\mathrm{sd}(\overline{X}) = \frac{\sigma}{\sqrt{n}} = \frac{20}{\sqrt{400}} = 1.0$

7.15 We first calculate the mean μ and standard deviation σ of the population that corresponds to X taking the values 3, 5, and 7, each having the same probability of occurring (namely 1/3).

x	$f(x)$	$xf(x)$	$x^2 f(x)$
3	1/3	3/3	9/3
5	1/3	5/3	25/3
7	1/3	7/3	49/3
Total	1	15/3	83/3

Using the values in the table, we have the following:

$$\mu = \sum x f(x) = \tfrac{15}{3} = 5$$

$$\sigma^2 = E(X^2) - \mu^2 = \sum x^2 f(x) - \mu^2 = \tfrac{83}{3} - 5^2 = \tfrac{8}{3}, \quad \text{so that } \sigma = \sqrt{\tfrac{8}{3}}$$

For $n = 2$, we know that the mean and standard deviation of the sampling distribution of $\overline{X}$ must be as follows:

$$E(\overline{X}) = \mu = 5$$

$$sd(\overline{X}) = \frac{\sigma}{\sqrt{2}} = \frac{\sqrt{\frac{8}{3}}}{\sqrt{2}} = \sqrt{\frac{8}{6}} = \sqrt{\frac{4}{3}}$$

We verify these by actually calculating the distribution of $\overline{X}$:

$\overline{x}$	$f(\overline{x})$	$\overline{x} f(\overline{x})$	$\overline{x}^2 f(\overline{x})$
3	1/9	3/9	9/9
4	2/9	8/9	32/9
5	3/9	15/9	75/9
6	2/9	12/9	72/9
7	1/9	7/9	49/9
Total	1	45/9	237/9

Using the values in the table, we have the following (which do indeed confirm the above assertion):

$$E(\overline{X}) = \sum \overline{x} f(\overline{x}) = \frac{45}{9} = 5$$

$$Var(\overline{X}) = E(\overline{X}^2) - (E(\overline{X}))^2 = \sum \overline{x}^2 f(\overline{x}) - (E(\overline{X}))^2 = \frac{237}{9} - 5^2 = \frac{12}{9} = \frac{4}{3},$$

$$sd(\overline{X}) = \sqrt{\frac{4}{3}}$$

7.17 (a) All possible samples (x_1, x_2) and the corresponding $\overline{x}$ values and probabilities are tabulated below. By independence, $P(x_1, x_2) = P(x_1) \cdot P(x_2)$, where the values of $P(1)$, $P(2)$, and $P(3)$ are given in the distribution of X. So, for instance, $P(1,1) = P(1) \cdot P(1) = (0.2)(0.2) = 0.04$.

(x_1, x_2)	(1,1)	(1,2)	(1,3)	(2,1)	(2,2)	(2,3)	(3,1)	(3,2)	(3,3)
Probability of (x_1, x_2)	0.04	0.12	0.04	0.12	0.36	0.12	0.04	0.12	0.04
$\overline{x} = \frac{x_1 + x_2}{2}$	1	1.5	2	1.5	2	2.5	2	2.5	3

The sampling distribution of $\overline{X}$ is obtained by listing the distinct values of $\overline{X}$ along with the corresponding probabilities, as follows:

$\overline{x}$	Probability $f(\overline{x})$
1	0.04
1.5	0.24
2	0.44
2.5	0.24
3	0.04
Total	1

(b) $E(\overline{X}) = \mu = \sum xf(x) = 1(0.2) + 2(0.6) + 3(0.2) = 2.0$. This is true for <u>any</u> sample size n.

(c) For $n = 36$, $E(\overline{X}) = 2.0$ (as mentioned in part (b)). Also, $sd(\overline{X}) = \frac{\sigma}{\sqrt{36}}$, where σ is the *population* standard deviation, which we calculate below.

x	$f(x)$	$xf(x)$	$x^2 f(x)$
1	0.2	0.2	0.2
2	0.6	1.2	2.4
3	0.2	0.6	1.8
Total	1	2.0	4.4

$\sigma^2 = E(X^2) - \mu^2 = \sum x^2 f(x) - \mu^2 = 4.4 - 2^2 = 0.4$, so that $\sigma = \sqrt{0.4} = 0.6325$

Thus, $sd(\overline{X}) = \frac{\sigma}{\sqrt{36}} = \frac{0.6325}{\sqrt{36}} = 0.1054$.

7.19 (a) $E(\overline{X}) = \mu = 37$

(b) $sd(\overline{X}) = \frac{\sigma}{\sqrt{n}} = \frac{6}{\sqrt{6}} = 2.449$

(c) Since the population distribution is normal, the sample mean $\overline{X}$ has normal distribution with mean 37 and standard deviation 2.449.

7.21 Denote X = weight of a package. We are given that X is normal with mean 8.1 and standard deviation 0.1.

(a) We convert to the standard normal to obtain
$$P[X < 8] = P[\frac{X - 8.1}{0.1} < \frac{8 - 8.1}{0.1}] = P[Z < -1] = 0.1587.$$
Hence, about 16% of the packages weigh less than the labeled amount.

(b) Let X_1 and X_2 denote the weight of two randomly chosen packages. Observe that:
$$E(\overline{X}) = 8.1$$
$$sd(\overline{X}) = \frac{0.1}{\sqrt{2}} = 0.0707$$
Hence, $\overline{X} = \frac{X_1 + X_2}{2}$ is normal with mean 8.1 and standard deviation 0.0707.

(c) Again, we convert to the standard normal (using part (b)) to obtain
$$P[\overline{X} < 8] = P[\frac{\overline{X} - 8.1}{0.0707} < \frac{8 - 8.1}{0.0707}] = P[Z < -1.414] = 0.0786.$$
Hence, there is about an 8% chance that the average weight of two packages will be less than the labeled amount of 8 ounces.

7.23 (a) We have $E(\overline{X}) = \mu = 31,000$ and $\text{sd}(\overline{X}) = \frac{\sigma}{\sqrt{n}} = \frac{5000}{\sqrt{100}} = 500$. Since $n = 100$ is

large, the central limit theorem ensures that the distribution of $\overline{X}$ is
approximately normal with mean and standard deviation as calculated above.

(b) The standardized variable is $Z = \dfrac{\overline{X} - 31,000}{500}$. As such, we have

$$P[\overline{X} > 31,500] = P[Z > \frac{31,500 - 31,000}{500}] = P[Z > 1] = 0.1587.$$

7.25 The population of fry has mean $\mu = 3.4$ and standard deviation $\sigma = 0.8$, so that

$$E(\overline{X}) = \mu = 3.4 \text{ and } \text{sd}(\overline{X}) = \frac{\sigma}{\sqrt{n}} = \frac{0.8}{\sqrt{36}} = 0.1333$$

and the standardized variable is $Z = \dfrac{\overline{X} - 3.4}{0.1333}$.

(a) $P[\overline{X} < 3.2] = P[Z < \frac{3.2 - 3.4}{0.1333}] = P[Z < -1.5] = 0.0668$

(b) Those caught in the net may be slower, less active fish, or even the less healthy
ones. Consequently, they may tend to be on the smaller side of the distribution.

7.27 We have $E(\overline{X}) = \mu = 34.5$ and $\text{sd}(\overline{X}) = \frac{\sigma}{\sqrt{n}} = \frac{1.3}{\sqrt{6}} = 0.5307$, and the standardized

variable is $Z = \dfrac{\overline{X} - 34.5}{0.5307}$. As such, we have

$$P[34.1 < \overline{X} < 35.2] = P[\frac{34.1 - 34.5}{0.5307} < Z < \frac{35.2 - 34.5}{0.5307}]$$
$$= P[-.0754 < Z < 1.319]$$
$$= P[Z < 1.319] - P[Z < -0.754] = 0.906 - 0.225 = 0.681.$$

7.29 (a) By column, the medians are as follows:

6	4	4	9	6	6	6
7	4	6	4	4	1	6
4	4	4	5	7	6	5
6	4	7	4	6	7	1
4	6	4	2	5	9	5

4	2	4	7	5	3	8
4	2	2	2	4	7	2
7	5	5	4	5	2	4
6	6	5	4	4	4	3
6	2	4	5	4	4	2

5	3	6	4	1	7
3	5	4	4	5	3
5	8	4	6	6	4
6	4	4	8	3	3
5	5	2	7	5	6

(b) & (c) Histograms are given below. The mean has smaller variance.

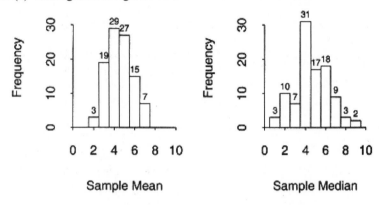

Frequency Chart:

x	1	2	3	4	5	6	7	8	9
frequency	3	10	7	31	17	18	9	3	2

7.31 (a) and (b) We use the following tabulated values to calculate the mean and standard deviation:

x	$f(x)$	$xf(x)$	$x^2 f(x)$
0	0.4	0	0
1	0.3	0.3	0.3
2	0.1	0.2	0.4
3	0.2	0.6	1.8
Total	1	1.1	1.8

Consequently, $E(X) = 1.1$ and $sd(X) = \sqrt{2.5 - (1.1)^2} = \sqrt{1.29} = 1.136$.

(c) Let X_1 be the number of complaints on the first day and X_2 the number of complaints on the second day. The total number of complaints is the sum $X_1 + X_2$. In order to determine the probability distribution of the sum, we make use of the fact that X_1 and X_2 are independent. For instance, the outcome (1,2), the total is $1 + 2 = 3$ and, by independence, the probability of the outcome (1,2) is the product $(0.3)(0.1) = 0.03$. All possible samples (x_1, x_2), along with the corresponding totals and probabilities, are tabulated below:

(x_1, x_2)	(0,0)	(0,1)	(0,2)	(0,3)	(1,0)	(1,1)	(1,2)	(1,3)
Probability of (x_1, x_2)	0.16	0.12	0.04	0.08	0.12	0.09	0.03	0.06
$x_1 + x_2$	0	1	2	3	1	2	3	4

(x_1, x_2)	(2,0)	(2,1)	(2,2)	(2,3)	(3,0)	(3,1)	(3,2)	(3,3)
Probability of (x_1, x_2)	0.04	0.03	0.01	0.02	0.08	0.06	0.02	0.04
$x_1 + x_2$	2	3	4	5	3	4	5	6

The sampling distribution of $X_1 + X_2$ is obtained by listing the distinct values of $X_1 + X_2$ along with the corresponding probabilities, as follows:

$x_1 + x_2$	Probability
0	0.16
1	0.24
2	0.17
3	0.22
4	0.13
5	0.04
6	0.04
Total	1

(d) Note that if the number of complaints is more than 125, the sample mean will be greater than $\frac{125}{90}$. Since the sample size $n = 90$ is large, we approximate the distribution of $\overline{X}$ by a normal distribution with mean 1.1 and standard deviation $\frac{\sigma}{\sqrt{n}} = \frac{1.136}{\sqrt{90}} = 0.1197$. Indeed, we have

$$P[\overline{X} > \tfrac{125}{90}] = P[Z > \frac{\frac{125}{90} - 1.1}{0.1197}] = P[Z > 2.413] = 0.0079.$$

7.33 (a) The histogram is given below:

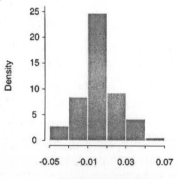

Two Month Differences

(b) There is more variability in the differences two months apart than in the one-month differences. A computer calculation gives the standard deviation 0.0197 versus 0.0128 for the one-month differences.

7.35 (a) By column, we record here the value of R for each possible sample listed in Exercise 7.34(a).

0	2	4	6
2	0	2	4
4	2	0	2
6	4	2	0

(b) For the sampling distribution of R, we list the distinct values along with the corresponding probabilities:

Value of R	Probability
0	4/16
2	6/16
4	4/16
6	2/16
Total	1

7.37 (a) All possible samples (x_1, x_2) and the corresponding $\bar{x}$ values and probabilities are tabulated below. By independence, $P(x_1, x_2) = P(x_1) \cdot P(x_2)$, where the values of $P(0)$, $P(1)$, and $P(2)$ are given in the distribution of X. So, for instance, $P(0,0) = P(0) \cdot P(0) = (0.3)(0.3) = 0.09$.

(x_1, x_2)	(0,0)	(0,1)	(0,2)	(1,0)	(1,1)	(1,2)	(2,0)	(2,1)	(2,2)
Probability of (x_1, x_2)	0.09	0.12	0.09	0.12	0.16	0.12	0.09	0.12	0.09
$\bar{x} = \frac{x_1 + x_2}{2}$	0	0.5	1	0.5	1	1.5	1	1.5	2

The sampling distribution of $\bar{X}$ is obtained by listing the distinct values of $\bar{X}$ along with the corresponding probabilities, as follows:

$\bar{x}$	Probability $f(\bar{x})$
0	0.09
0.5	0.24
1	0.34
1.5	0.24
2	0.09
Total	1

(b) $E(\overline{X}) = \mu = \sum xf(x) = 0(0.3) + 1(0.4) + 3(0.3) = 1.0$. This is true for <u>any</u> sample size n.

(c) For $n = 81$, $E(\overline{X}) = 1.0$ (as mentioned in part (b)). Also, $\text{sd}(\overline{X}) = \frac{\sigma}{\sqrt{81}}$, where σ is the *population* standard deviation, which we calculate below.

x	$f(x)$	$xf(x)$	$x^2 f(x)$
0	0.3	0	0
1	0.4	0.4	0.4
2	0.3	0.6	1.2
Total	1	1.0	1.6

$\sigma^2 = E(X^2) - \mu^2 = \sum x^2 f(x) - \mu^2 = 1.6 - 1^2 = 0.6$, so that $\sigma = \sqrt{0.6} = 0.7746$

Thus, $\text{sd}(\overline{X}) = \frac{\sigma}{\sqrt{81}} = \frac{0.7746}{\sqrt{81}} = 0.086$.

7.39 We have $\mu = 60$, $\sigma = 10$, and $n = 9$.

(a) $E(\overline{X}) = \mu = 60$, $\quad \text{sd}(\overline{X}) = \frac{\sigma}{\sqrt{n}} = \frac{10}{\sqrt{9}} = \frac{10}{3}$

(b) Since the population distribution is normal, the sample mean $\overline{X}$ has normal distribution with mean and standard deviation given in part (a).

(c) The standardized variable is $Z = \dfrac{\overline{X} - 60}{10/3}$. As such, we have

$$P[56 < \overline{X} < 64] = P[\frac{56 - 60}{10/3} < Z < \frac{64 - 60}{10/3}] = P[-1.2 < Z < 1.2]$$
$$= P[Z < 1.2] - P[Z < -1.2] = 0.8849 - 0.1151 = 0.7698.$$

7.41 We have $\mu = 12.1$, $\sigma = 3.2$, and $n = 9$.

(a) Since the population is normal, the distribution of $\overline{X}$ is normal with mean $= \mu = 12.1$ and $\text{sd} = \frac{\sigma}{\sqrt{n}} = \frac{3.2}{\sqrt{9}} = \frac{3.2}{3}$.

(b) The standardized normal variable is $Z = \dfrac{\overline{X} - 12.1}{3.2/3}$. As such, observe that

$$P[\overline{X} < 10] = P[Z < \frac{10 - 12.1}{3.2/3}] = P[Z < -1.97] = 0.0244.$$

(c) The pebble size X is normally distributed with mean 12.1 and standard deviation 3.2, so $Z = \dfrac{X - 12.1}{3.2}$ is $N(0,1)$. As such, observe that

$$P[X < 10] = P[Z < \frac{10 - 12.1}{3.2}] = P[Z < -0.656] = 0.256.$$

So, about 26% of the pebbles are of size smaller than 10.

7.43 We have $\mu = 54$, $\sigma = 8$, and $n = 81$.

(a) $E(\overline{X}) = \mu = 54$ and $sd(\overline{X}) = \frac{\sigma}{\sqrt{n}} = \frac{8}{\sqrt{81}} = 0.889$.

(b) Since $n = 81$ is large, the central limit theorem ensures that the distribution of $\overline{X}$ is approximately normal with mean and standard deviation as calculated in part (a).

7.45 (a) Since $n = 45$ is large, the central limit theorem ensures that the distribution of $\overline{X}$ is approximately normal with mean $\mu = 55$ and standard deviation

$= \frac{\sigma}{\sqrt{n}} = \frac{7}{\sqrt{45}} = 1.0435$. As such,

$$P[54 < \overline{X} < 56] = P[\frac{54-55}{1.0435} < Z < \frac{56-55}{1.0435}] = P[-0.958 < Z < 0.958]$$
$$= P[Z < 0.958] - P[Z < -0.958] = 0.8310 - 0.1690 = 0.6621.$$

(b) For a standard normal variable Z, we know that $P[-1.96 < Z < 1.96] = 0.95$.

The required interval for $\overline{X}$ is

$55 \pm 1.96 \big(sd(\overline{X})\big) = 55 \pm 1.96(1.0435) = 55 \pm 2.05$ or (52.95, 57.05).

7.47 (a) & (b) We calculate the population mean and standard deviation using the tabulated values below:

x	$f(x)$	$xf(x)$	$x^2 f(x)$
0	0.5	0	0
1	0.3	0.3	0.3
2	0.2	0.4	0.8
Total	1	0.7	1.1

Using the values in the table, we have the following:

$\mu = \sum x f(x) = 0.7$

$\sigma^2 = E(X^2) - \mu^2 = \sum x^2 f(x) - \mu^2 = 1.1 - (0.7)^2 = 0.61$, so that $\sigma = \sqrt{0.61} = 0.781$

(c) Let X_1 be the number sold the next day and X_2 the number sold the day after next. The total number sold is the sum $X_1 + X_2$. In order to determine the probability distribution of the sum, we make use of the fact that X_1 and X_2 are independent. For instance, the outcome (2,1), the total is $2+1 = 3$ and, by independence, the probability of the outcome (2,1) is the product $(0.2)(0.3) = 0.06$. All possible samples (x_1, x_2), along with the corresponding totals and probabilities, are tabulated below:

(x_1, x_2)	(0,0)	(0,1)	(0,2)	(1,0)	(1,1)	(1,2)	(2,0)	(2,1)	(2,2)
Probability of (x_1, x_2)	0.25	0.15	0.10	0.15	0.09	0.06	0.10	0.06	0.04
$x_1 + x_2$	0	1	2	1	2	3	2	3	4

The sampling distribution of $X_1 + X_2$ is obtained by listing the distinct values of $X_1 + X_2$ along with the corresponding probabilities, as follows:

$x_1 + x_2$	Probability
0	0.25
1	0.30
2	0.29
3	0.12
4	0.04
Total	1

(d) The event that at least 53 kayaks are sold is the same event that $\overline{X} > \frac{53}{64}$. Since the sample size $n = 64$ is large, we approximate the distribution of $\overline{X}$ by a normal distribution with mean 0.7 and standard deviation $\frac{\sigma}{\sqrt{n}} = \frac{0.781}{\sqrt{64}} = 0.09763$.

Indeed, we have $P[\overline{X} > \frac{53}{64}] = P[Z > \frac{\frac{53}{64} - 0.7}{0.09763}] = P[Z > 1.312] = 0.0951$.

(e) We must have the z-value 1.645. If k is the required number, then

$$\frac{\frac{k}{64} - 0.7}{0.09763} = 1.645,$$

so that $k = 64(0.7 + 1.645(0.09763)) = 55.1$. Hence, 56 kayaks must be ordered.

7.49 (a) Let X be the amount in a single bottle. The distribution of X is normal, so that the desired probability is

$$P[X < 299] = P[\frac{X - 302}{2} < \frac{299 - 302}{2}] = P[Z < -1.5] = 0.0668.$$

(b) Since $P[Z > 1.645] = 0.0500$, we have

$$\frac{v - 302}{2} = 1.645 \quad \text{or} \quad v = 302 + 2(1.645) = 305.29 \text{ ml}.$$

(c) Since the population is normal, $\overline{X}$ has a normal distribution with mean 302 and standard deviation $\frac{\sigma}{\sqrt{n}} = \frac{2}{\sqrt{2}} = \sqrt{2}$. Indeed, we have

$$P[\overline{X} < 299] = P[Z < \frac{299 - 302}{\sqrt{2}}] = P[Z < -2.121] = 0.017.$$

(d) From part (c), each package has probability of 0.017 of containing less than 299 ml. Let Y be the number out of 2 packages that contain less than 299 ml. Then, Y has a binomial distribution with $n = 2$ and $p = 0.017$. We are interested in the event $Y = 1$. By the formula for the binomial distribution, we see that the probability of Y occurring is: $2(0.017)(1 - 0.017) = 0.0334$.

Chapter 8

DRAWING INFERENCES FROM LARGE SAMPLES

8.1 (a) (i) $S.E.= \frac{\sigma}{\sqrt{n}} = \frac{22}{\sqrt{152}} = 1.784$

      (ii) 95% error margin is $z_{0.025} \frac{\sigma}{\sqrt{n}} = (1.96)(1.784) = 3.50$.

  (b) (i) $S.E.= \frac{\sigma}{\sqrt{n}} = \frac{8.2}{\sqrt{85}} = 0.889$

      (ii) 99% error margin is $z_{0.005} \frac{\sigma}{\sqrt{n}} = (2.58)(0.889) = 2.29$.

  (c) (i) $S.E.= \frac{\sigma}{\sqrt{n}} = \frac{56}{\sqrt{295}} = 3.260$

      (ii) 92% error margin is $z_{0.04} \frac{\sigma}{\sqrt{n}} = (1.75)(3.260) = 5.71$.

8.3 Point estimate is $\bar{x}$. Estimated standard error is $\frac{s}{\sqrt{n}}$, where $s = \sqrt{\dfrac{\sum(x_i - \bar{x})^2}{n-1}}$.

  (a) $\bar{x} = \dfrac{752}{70} = 10.74, \quad s^2 = \dfrac{235}{69} = 3.406, \quad \dfrac{s}{\sqrt{70}} = 0.221$

  (b) $\bar{x} = \dfrac{2653}{90} = 29.48, \quad s^2 = \dfrac{546}{89} = 6.135, \quad \dfrac{s}{\sqrt{90}} = 0.261$

  (c) $\bar{x} = \dfrac{3985}{160} = 24.91, \quad s^2 = \dfrac{745}{159} = 4.686, \quad \dfrac{s}{\sqrt{160}} = 0.171$

8.5 Estimated mean weakly amount contested is $\bar{x} = \$75.43$.

    Estimated standard error is $\dfrac{s}{\sqrt{n}} = \dfrac{24.73}{\sqrt{50}} = 3.497$.

    90% error margin is $1.645\dfrac{s}{\sqrt{n}} = 1.645(3.497) = \$5.75$.

8.7   (a) Given that the 95% error margin is $1.96\dfrac{s}{\sqrt{n}} = 4.2$, we know that the estimated

      standard error is $\dfrac{s}{\sqrt{n}} = \dfrac{4.2}{1.96} = 2.143$.

    (b) 90% error margin is $1.645\dfrac{s}{\sqrt{n}} = 1.645(2.143) = 3.525$.

8.9   Recall that the 95% error margin is $1.96\dfrac{\sigma}{\sqrt{n}}$.

    (a) In order that the 95% error margin be equal to $\frac{1}{8}\sigma$, we must have

$$1.96\frac{\sigma}{\sqrt{n}} = \tfrac{1}{8}\sigma,$$

    so that solving for n yields $n = (1.96 \times 8)^2 = 245.9$. So, the required sample size
    is $n = 246$.

    (b) In order that the 95% error margin be equal to 0.15σ, we must have

$$1.96\frac{\sigma}{\sqrt{n}} = 0.15\sigma,$$

    so that solving for n yields $n = \left(\dfrac{1.96}{0.15}\right)^2 = 170.7$. So, the required sample size is

    $n = 171$.

8.11 We have $d = 5.00$, $\sigma = 25$, and $z_{\alpha/2} = z_{0.01} = 2.33$. Hence,

$$\left[\frac{2.33(25)}{5.00}\right]^2 = 135.7.$$

So, the required sample size is $n = 136$.

8.13 We have $d = 0.5$, $\sigma = 2.5$, and $z_{\alpha/2} = z_{0.05} = 1.645$. Hence,

$$\left[\frac{1.645(2.5)}{0.5}\right]^2 = 67.6.$$

So, the required sample size is $n = 68$.

8.15 Observe that $1 - \alpha = 0.90$ implies $\alpha = 0.10$, so that $z_{\alpha/2} = z_{0.05} = 1.645$. A 90%
    confidence interval for μ is calculated as follows:

$$\bar{x} \pm 1.645\frac{s}{\sqrt{n}} = 81.3 \pm 1.645\frac{5.8}{\sqrt{63}} = 81.3 \pm 1.20 \text{ or } (80.10, 82.50).$$

8.17 95% of 365, namely 347, confidence intervals are expected to cover the true means. Before the data are obtained each day, the probability of covering the population mean is 0.95. By the long run frequency interpretation of probability, approximately 95% will cover.

8.19 For large n, a 95% confidence interval for μ is given by $\overline{X} \pm z_{0.025}\dfrac{S}{\sqrt{n}}$. Using

$z_{0.025} = 1.96$, $n = 35$, and the summary statistics $\overline{x} = 3.20$ grams, $s = 3.8$ grams, the 95% confidence interval for μ is given by

$$30.2 \pm 1.96\frac{3.8}{\sqrt{35}} \;=\; 30.2 \pm 1.26 \;\; \text{or} \;\; (28.94, 31.46) \text{ grams.}$$

8.21 For large n, a 99% confidence interval for μ is given by $\overline{X} \pm z_{0.005}\dfrac{S}{\sqrt{n}}$. Using

$z_{0.005} = 2.58$, $n = 120$, and the summary statistics $\overline{x} = 18.3$ days, $s = 5.2$ days, the 99% confidence interval for μ (true mean survival time) is given by

$$18.3 \pm 2.58\frac{5.2}{\sqrt{120}} \;=\; 18.3 \pm 1.2 \;\; \text{or} \;\; (17.1, 19.5) \text{ days.}$$

8.23 For large n, a 95% confidence interval for μ is given by $\overline{X} \pm z_{0.025}\dfrac{S}{\sqrt{n}}$. Using

$z_{0.025} = 1.96$, $n = 70$, and the summary statistics $\overline{x} = 493$, $s = 72$, the 95% confidence interval for μ (population mean math score) is given by

$$493 \pm 1.96\frac{72}{\sqrt{70}} \;=\; 493 \pm 16.9 \;\; \text{or} \;\; (476.1, 509.9).$$

8.25 For large n, a 90% confidence interval for μ is given by $\overline{X} \pm z_{0.05}\dfrac{S}{\sqrt{n}}$. Using

$z_{0.05} = 1.645$, $n = 140$, and the summary statistics $\overline{x} = 8.6$ miles, $s = 4.3$ miles, the 90% confidence interval for μ (true mean commuting distance) is given by

$$8.6 \pm 1.645\frac{4.3}{\sqrt{140}} \;=\; 8.6 \pm 0.60 \;\; \text{or} \;\; (8.00, 9.20) \text{ miles.}$$

8.27 For large n, a 95% confidence interval for μ is given by $\overline{X} \pm z_{0.025}\dfrac{S}{\sqrt{n}}$. Using

$z_{0.025} = 1.96$, $n = 50$, and the summary statistics $\overline{x} = 75.43$ dollars, $s = 24.73$ dollars, the 95% confidence interval for μ (true mean amount contested) is given by

$$75.43 \pm 1.96\frac{24.73}{\sqrt{50}} \;=\; 75.43 \pm 6.85 \;\; \text{or} \;\; (68.58, 82.28) \text{ dollars.}$$

8.29 For large n, a 95% confidence interval for μ is given by $\overline{X} \pm z_{0.025}\dfrac{S}{\sqrt{n}}$. Using

$z_{0.025} = 1.96$, $n = 147$, and the summary statistics $\overline{x} = 0.0021$, $s = 0.0159$, the 95% confidence interval for μ is given by

$$0.0021 \pm 1.96\frac{0.0159}{\sqrt{147}} = 0.0021 \pm 0.0026 \text{ or } (-0.0005,\ 0.0047\,).$$

8.31 For large n, a 99% confidence interval for μ is given by $\overline{X} \pm z_{0.005}\dfrac{S}{\sqrt{n}}$. Using

$z_{0.005} = 2.58$, $n = 40$, and the summary statistics $\overline{x} = 1.715$ centimeters, $s = 0.475$ centimeters, the 99% confidence interval for μ is given by

$$1.715 \pm 2.58\frac{0.475}{\sqrt{40}} = 1.715 \pm 0.194 \text{ or } (1.521,\ 1.909) \text{ centimeters.}$$

8.33 (a) Uncertain since the true mean growth (a <u>population</u> parameter) is not known. Refer to the discussion on pages 301 – 302 of the text for a detailed explanation.

(b) The very definition of a confidence interval ensures this statement is true – see property 2 on page 302 of the text.

8.35 (a) For large n, a 95% confidence interval for μ is given by $\overline{X} \pm z_{0.025}\dfrac{S}{\sqrt{n}}$. Using

$z_{0.025} = 1.96$, $n = 40$, and the summary statistics $\overline{x} = 3.56$, $s = 0.05$, the 95% confidence interval for μ (true mean amount of PCBs in the soil) is given by

$$3.56 \pm 1.96\frac{0.05}{\sqrt{40}} = 3.56 \pm 0.155 \text{ or } (3.405,\ 3.715) \text{ PCBs.}$$

(b) The sample mean is $\overline{x} = 3.56$ and <u>does</u> indeed lie in this interval. In fact, it is the midpoint of the interval. (This is <u>always</u> true for the *sample* mean.)

(c) Uncertain since the true mean growth (a <u>population</u> parameter) is not known. Refer to the discussion on pages 301 – 302 of the text for a detailed explanation.

(d) No. This is an incorrect interpretation of a confidence interval. Refer to pages 301 – 302 of the text for a correct interpretation.

8.37 The alternative hypothesis H_1 is the assertion that is to be established; its opposite is the null hypothesis H_0.

(a) Let μ denote the population mean time, in days, to pay a claim. The hypotheses are: $H_0 : \mu = 14$, $H_1 : \mu < 14$

(b) Let μ denote the population mean amount spent, in dollars. The hypotheses are: $H_0 : \mu = 2.50$, $H_1 : \mu > 2.50$

(c) Let μ denote the population mean hospital bill, in dollars. The hypotheses are: $H_0: \mu=3000, \quad H_1: \mu<3000$

(d) Let μ denote the population mean time between purchases, in days. The hypotheses are: $H_0: \mu=60, \quad H_1: \mu \neq 60$

8.39 Retaining H_0 is a correct decision if H_0 is true, while it is an incorrect decision if H_1 is true. A Type II error is incurred when such an incorrect decision is made.

(a) Correct decision if $\mu=14$ and incorrect decision if $\mu<14$ (Type II error).

(b) Correct decision if $\mu=2.50$ and incorrect decision if $\mu>2.50$ (Type II error).

(c) Correct decision if $\mu=3000$ and incorrect decision if $\mu<3000$ (Type II error).

(d) Correct decision if $\mu=60$ and incorrect decision if $\mu \neq 60$ (Type II error).

8.41 (b) (i) $H_0: \mu=0.15, \quad H_1: \mu<0.15$

(ii) $Z=\dfrac{\overline{X}-0.15}{0.085/\sqrt{125}}$

(iii) $R: Z \leq -1.96$

(c) (i) $H_0: \mu=80, \quad H_1: \mu \neq 80$

(ii) $Z=\dfrac{\overline{X}-80}{8.6/\sqrt{38}}$

(iii) $R:|Z| \geq 2.58$

(d) (i) $H_0: \mu=0, \quad H_1: \mu \neq 0$

(ii) $Z=\dfrac{\overline{X}-0}{1.23/\sqrt{40}}$

(iii) $R:|Z| \geq 1.88$

8.43 Since the claim is that $\mu>30$, we formulate the hypotheses:

$$H_0: \mu=30, \quad H_1: \mu>30$$

With $n=70$ and $\sigma=5.6$, the test statistic (in standardized form) is $Z=\dfrac{\overline{X}-30}{5.6/\sqrt{70}}$.

(a) $\overline{X} \geq 31.31$ is equivalent to $Z \geq \dfrac{31.31-30}{5.6/\sqrt{70}}=1.96$, or more simply $Z \geq 1.96$.

Since $P[Z \geq 1.96]=0.025$, the level of significance is $\alpha=0.025$.

(b) Since $P[Z \geq 1.645]=0.05$, the rejection region $R: Z \geq 1.645$ has the level of significance $\alpha=0.05$. Now, since $Z=\dfrac{\overline{X}-30}{5.6/\sqrt{70}}$, we observe that $Z \geq 1.645$

is equivalent to $\overline{X} \geq 30+(1.645)\dfrac{5.6}{\sqrt{70}}=31.10$. Therefore, $c=31.10$.

8.45 The test statistic remains the same, but this time the rejection region is two-sided. With $\alpha = 0.05$, we have $z_{\alpha/2} = z_{0.025} = 1.96$ and the two-sided rejection region is $R : |Z| \geq 1.96$. Note that the observed value $|z| = 1.88$ is not in this rejection region. Hence, H_0 is not rejected at $\alpha = 0.05$. Furthermore, the p-value is $2P[Z \leq -1.88] = 2(0.0301) = 0.0602$. The rejection region and p-value are illustrated below (in that order):

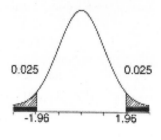 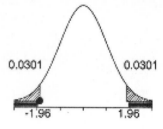

8.47 (a) Reject H_0 since the p-value is 0.005, which is less than 0.01.
       (b) We could have made a Type I error if H_0 happened to be true.
       (c) Prior to sampling, the probability of making a Type I error is 0.01.
       (d) If we took more and more samples and ran the same test of hypothesis, about 1% of the time we would make a Type I error (i.e., reject a true H_0).
       (e) The p-value is 0.005, as reported on the Minitab output.

8.49 Since the claim is that $\mu > 3.5$, we formulate the hypotheses:
$$H_0 : \mu = 3.5, \quad H_1 : \mu > 3.5$$

Now, to run the test with sample size $n = 40$, we use the test statistic $Z = \dfrac{\overline{X} - 3.5}{S / \sqrt{40}}$.

Since H_1 is right-sided, the rejection region should have the form $R : Z \geq z_\alpha$. With $\alpha = 0.10$, the rejection region is $R : Z \geq 1.28$. The observed value of the test statistic is $z = \dfrac{3.8 - 3.5}{1.2 / \sqrt{40}} = 1.58$, which is in R. Therefore, we reject H_0 at $\alpha = 0.10$.

Furthermore, the p-value is $P[Z \geq 1.58] = 0.0571$. The rejection region and p-value are illustrated at the top of the next page (in that order):

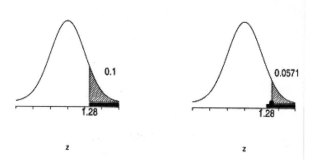

8.51 For each of the following, the sample size is $n = 36$ and the test statistic is

$$Z = \frac{\overline{X} - \mu_0}{S/\sqrt{36}}.$$

(a) For $\mu_0 = 74$, the test statistic is $Z = \dfrac{\overline{X} - 74}{S/\sqrt{36}}$.

Since H_1 is right-sided, the rejection region should have the form $R : Z \geq c$ (i.e., it should consist of large values of Z). The observed value of the test statistic is

$$z = \frac{80.4 - 74}{16.2/\sqrt{36}} = 2.37.$$

Observe that the p-value is $P[Z \geq 2.37] = 0.0089$. This extremely small p-value signifies a strong rejection of H_0.

(b) For $\mu_0 = 85$, the test statistic is $Z = \dfrac{\overline{X} - 85}{S/\sqrt{36}}$.

Since H_1 is left-sided, the rejection region should have the form $R : Z \leq c$ (i.e., it should consist of small values of Z). The observed value of the test statistic is

$$z = \frac{80.4 - 85}{16.2/\sqrt{36}} = -1.70.$$

Observe that the p-value is $P[Z \leq -1.70] = 0.0446$. The null hypothesis H_0 would be rejected for α as small as 0.0446, which is about 4%. There is fairly strong evidence for the rejection of H_0.

(c) For $\mu_0 = 76$, the test statistic is $Z = \dfrac{\overline{X} - 76}{S/\sqrt{36}}$.

Since H_1 is two-sided, the rejection region should have the form $R : |Z| \geq c$ (i.e., it should consist of large values of $|Z|$). The observed value of the test statistic is

$$z = \frac{80.4 - 76}{16.2 / \sqrt{36}} = 1.63.$$

Observe that the p-value is

$$P[|Z| \geq 1.63] = 2P[Z \leq -1.63] = 2(0.0516) = 0.1032.$$

The smallest α at which the null hypothesis H_0 would be rejected for 0.1032, which is about 10%. Since this is not very small, the support for H_1 is not strong.

8.53 Let μ denote the population mean hold time (in minutes). Since the claim is that $\mu > 3.0$, we formulate the following hypotheses: $H_0 : \mu = 3.0, \quad H_1 : \mu > 3.0$

Now, to run the test with sample size $n = 75$, we use the test statistic $Z = \dfrac{\overline{X} - 3.0}{S / \sqrt{75}}$.

Since H_1 is right-sided, the rejection region should have the form $R : Z \geq z_\alpha$. With $\alpha = 0.05$, the rejection region is $R : Z \geq 1.645$. The observed value of the test statistic is

$$z = \frac{3.4 - 3.0}{2.4 / \sqrt{75}} = 1.44,$$

which is not in R. Therefore, we do not reject H_0 at $\alpha = 0.05$. Furthermore, the p-value is $P[Z \geq 1.44] = 0.0749$. Since this is not small, support for the claim is weak.

8.55 Let μ denote the true mean BOD. Since the intent is to establish that μ is different from 3000, we formulate the following hypotheses:

$$H_0 : \mu = 3000, \quad H_1 : \mu \neq 3000$$

Now, to run the test with sample size $n = 43$, we use the test statistic $Z = \dfrac{\overline{X} - 3000}{S / \sqrt{43}}$

Since H_1 is two-sided, the rejection region should have the form $R : |Z| \geq z_{\alpha/2}$. With $\alpha = 0.05$, $z_{0.025} = 1.96$, so that the rejection region is $R : |Z| \geq 1.96$. The observed value of the test statistic is

$$z = \frac{3246 - 3000}{757 / \sqrt{43}} = 2.13,$$

which is in R. Therefore, we reject H_0 at $\alpha = 0.05$. This means that there is strong evidence that the BOD is significantly off the target.

8.57 (a)   Point estimate $\hat{p} = \frac{28}{50} = 0.56$

Estimated S.E. $= \sqrt{\dfrac{\hat{p}\hat{q}}{n}} = \sqrt{\dfrac{(0.56)(0.44)}{50}}$

So, the 95% error margin $= 1.96 \,(\text{Estimated S.E.}) = 1.96 \sqrt{\dfrac{(0.56)(0.44)}{50}} = 0.138$.

(b)   Point estimate $\hat{p} = \frac{75}{410} = 0.183$

Estimated S.E. $= \sqrt{\frac{\hat{p}\hat{q}}{n}} = \sqrt{\frac{(0.183)(0.817)}{410}}$

So, the 95% error margin $=1.96\,(\text{Estimated S.E.})=1.96\sqrt{\frac{(0.183)(0.817)}{410}} = 0.037$.

(c) Point estimate $\hat{p} = \frac{2001}{2500} = 0.800$

Estimated S.E. $= \sqrt{\frac{\hat{p}\hat{q}}{n}} = \sqrt{\frac{(0.800)(0.200)}{2500}}$

So, the 95% error margin $=$

$$1.96\,(\text{Estimated S.E.})=1.96\sqrt{\frac{(0.800)(0.200)}{2500}} = 0.016.$$

8.59 (a) We have $p = 0.3$, $q = 0.7$, $d = 0.03$, $\alpha = 0.10$. So, $z_{\alpha/2} = z_{0.05} = 1.645$. We

calculate $n = (0.3)(0.7)\left[\frac{1.645}{0.03}\right]^2 = 631.40$. So, the required sample size is 632.

(b) Since p is unknown, the conservative bound on n yields $\frac{1}{4}\left[\frac{1.645}{0.03}\right]^2 = 751.67$,

so that the required sample size in this instance is 752.

8.61 (a) The estimate of the market share is $\hat{p} = \frac{49}{78} = 0.6282$.

(b) Estimated S.E. $= \sqrt{\frac{\hat{p}\hat{q}}{n}} = \sqrt{\frac{(0.3718)(0.6282)}{78}} = 0.0547$

(c) We have $\hat{p} = \frac{49}{78} = 0.628$, so that $\hat{q} = 0.372$. We also are given that $\alpha = 0.05$

(so that $z_{\alpha/2} = z_{0.025} = 1.96$) and $n = 78$. Thus, the 95% confidence interval is

$$\hat{p} \pm z_{0.025}\sqrt{\frac{\hat{p}\hat{q}}{n}} = 0.628 \pm 1.96\sqrt{\frac{(0.372)(0.628)}{78}} \quad \text{or} \quad (0.521, 0.735).$$

8.63 (a) Denote $p_M =$ population proportion of ERS calls involving serious
mechanical problems. Since there were 849 calls involving serious
mechanical problems out of $n = 2927$ ERS calls, the point estimate of p_M is

$\hat{p}_M = \frac{849}{2927} = 0.290$. Also, the 95% error margin is $1.96\sqrt{\frac{(0.29)(0.71)}{2927}} = 0.016$.

(b) Denote $p_S =$ population proportion of ERS calls involving starting problems.
Since there were 1499 calls involving starting problems out of $n = 2927$ ERS
calls, the point estimate of p_S is $\hat{p}_S = \frac{1499}{2927} = 0.512$, and so
$\hat{q}_S = 1 - 0.512 = 0.488$. The 98% confidence interval for p_S is then given by

$$\hat{p}_S \pm 2.33\sqrt{\frac{\hat{p}_S\hat{q}_S}{n}} = 0.512 \pm 2.33\sqrt{\frac{(0.512)(0.488)}{2927}} = 0.512 \pm 0.0215$$

or $(0.491, 0.534)$.

8.65 (a) Let p be the proportion of students that hold a part time job. The hypotheses
are: $H_0 : p = 0.26, \quad H_1 : p > 0.26$

(b) Let p be the proportion of subscribers that have complaints against the cable
company. The hypotheses are: $H_0 : p = 0.13, \quad H_1 : p < 0.13$

(c) Let p be the proportion of subscribers that have complaints against the cable
company. The hypotheses are: $H_0 : p = 0.13, \quad H_1 : p > 0.13$

(d) Let p be the proportion of boards that would break under the standard load.
The hypotheses are: $H_0 : p = 0.05, \quad H_1 : p < 0.05$

8.67 (b) (i) $H_0 : p = 0.75, \quad H_1 : p > 0.75$

(ii) $Z = \dfrac{\hat{p} - 0.75}{\sqrt{\frac{(0.75)(0.25)}{228}}} = \dfrac{\hat{p} - 0.75}{0.0287}$

(iii) Since $\alpha = 0.02$, $z_{0.02} = 2.05$ and H_1 is right-sided, the rejection region is
given by $R : Z \geq 2.05$.

(c) (i) $H_0 : p = 0.60, \quad H_1 : p \neq 0.60$

(ii) $Z = \dfrac{\hat{p} - 0.60}{\sqrt{\frac{(0.60)(0.40)}{77}}} = \dfrac{\hat{p} - 0.60}{0.0558}$

(iii) Since $\alpha = 0.02$, $\alpha/2 = 0.01$, $z_{0.01} = 2.33$ and H_1 is two-sided, the rejection
region is given by $R : |Z| \geq z_{\alpha/2} = 2.33$.

(d) (i) $H_0 : p = 0.56, \quad H_1 : p < 0.56$

(ii) $Z = \dfrac{\hat{p} - 0.56}{\sqrt{\frac{(0.56)(0.44)}{86}}} = \dfrac{\hat{p} - 0.56}{0.0535}$

(iii) Since $\alpha = 0.10$, $z_{0.10} = 1.28$ and H_1 is left-sided, the rejection region is
given by $R : Z \leq -1.28$.

8.69 Let $\hat{p} = $ sample proportion.

(a) Test statistic: $Z = \dfrac{\hat{p} - 0.40}{\sqrt{\frac{(0.40)(0.60)}{n}}}$

Since $z_{0.10} = 1.28$ and H_1 is left-sided, the rejection region is $R : Z \leq -1.28$.

(b) Test statistic: $Z = \dfrac{\hat{p} - 0.60}{\sqrt{\frac{(0.60)(0.40)}{n}}}$

Since $\alpha = 0.10$, $z_{0.10/2} = 1.645$ and H_1 is two-sided, the rejection region is
$R : |Z| \geq 1.645$.

8.71 (a) Since the intent is to establish that $p < 0.5$, we formulate the following hypotheses: $H_0 : p = 0.5, \quad H_1 : p < 0.5$

   (b) The sample proportion of support is $\hat{p} = \frac{228}{500} = 0.456$. Since $\alpha = 0.05$, $z_{0.05} = 1.645$ and H_1 is left-sided, the rejection region is $R : Z \leq -1.645$. The test statistic is calculated to be $Z = \dfrac{\frac{228}{500} - 0.5}{\sqrt{\frac{(0.5)(0.5)}{500}}} = -1.97$, which is in R. Hence, H_0 is rejected at $\alpha = 0.05$ and the claim is supported. Furthermore, the p-value is $P[Z \leq -1.97] = 0.0224$, so that H_0 would still be rejected with α as small as 0.0224. The rejection region and p-value are illustrated below (in that order):

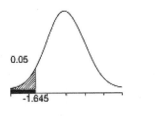
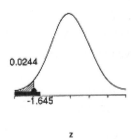

8.73 Let p denote the population proportion of high school graduates who enrolled in colleges. Since the intent is to establish that $p > 0.48$, we formulate the following hypotheses: $H_0 : p = 0.48, \quad H_1 : p > 0.48$

The sample proportion of support is $\hat{p} = 0.503$. Since H_1 is right-sided, the rejection region is of the form $R : Z \geq c$. The test statistic, based on a sample of size $n = 700$, is calculated to be $Z = \dfrac{\hat{p} - 0.48}{\sqrt{\frac{(0.48)(0.52)}{700}}} = \dfrac{0.503 - 0.48}{0.019} = 1.21$. The associated p-value to this observed z is $P[Z \geq 1.21] = 0.1131$, which is about 11.31% (this is illustrated below). So, we would need to take an α of 11.31% or higher in order to reject H_0.

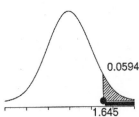

8.75 (a)   We have $n = 505$ and $\hat{p} = \frac{258}{505} = 0.5109$. So, a 90% confidence interval for p
is given by:

$$\hat{p} \pm z_{0.05} \sqrt{\frac{\hat{p}\hat{q}}{n}} = 0.5109 \pm 1.645 \sqrt{\frac{(0.5109)(0.4881)}{505}} = 0.5109 \pm 0.0366$$

or $(0.4743, 0.5475)$.

(b)   The unknown total number of customers who would rate the service as being
excellent is $8200p$. The lower endpoint of the 90% confidence interval for the
total is then $8200(0.4743) = 3889$, and the upper endpoint is $8200(0.5475) = 4490$.

8.77   We have $n = 42$ and $\hat{p} = \frac{30}{42} = 0.7143$. So, a 95% confidence interval for p
is given by:

$$\hat{p} \pm z_{0.025} \sqrt{\frac{\hat{p}\hat{q}}{n}} = 0.7143 \pm 1.96 \sqrt{\frac{(0.7143)(0.2857)}{42}} = 0.7143 \pm 0.1366$$

or $(0.58, 0.85)$. That is, $(58\%, 85\%)$.

8.79   Point estimate is $\bar{x} = \dfrac{\sum x_i}{n}$. Estimated standard error is $\frac{s}{\sqrt{n}}$, where $s = \sqrt{\dfrac{\sum (x_i - \bar{x})^2}{n-1}}$.

(a) $\bar{x} = \dfrac{752}{80} = 9.40$, $s^2 = \dfrac{345}{79} = 4.367$, $\text{S.E.} = \dfrac{s}{\sqrt{n}} = 0.234$

(b) $\bar{x} = \dfrac{1290}{169} = 7.63$, $s^2 = \dfrac{842}{168} = 5.012$, $\text{S.E.} = \dfrac{s}{\sqrt{n}} = 0.172$

8.81   Recall that the standard error of $\bar{X}$ is $\dfrac{\sigma}{\sqrt{n}}$, where n is the sample size.

(a) Since $\dfrac{1}{2}\dfrac{\sigma}{\sqrt{n}} = \dfrac{\sigma}{\sqrt{4n}}$, we need a sample of size $4n$. Therefore, we must increase
the sample size by a factor of 4.

(b) Since $\dfrac{1}{4}\dfrac{\sigma}{\sqrt{n}} = \dfrac{\sigma}{\sqrt{16n}}$, we need a sample of size $16n$. Therefore, we must
increase the sample size by a factor of 16.

8.83 (a)   The population mean μ is estimated by $\bar{x} = 126.9$.

Estimated $\text{S.E.} = \dfrac{s}{\sqrt{n}} = \dfrac{10.5}{\sqrt{55}} = 1.416$. So, the approximate 95.4% error margin
is given by 2 (Estimated S.E.) $= 2(1.416) \approx 2.8$,

(b)   Observe that $1 - \alpha = 0.90$ implies $\alpha = 0.10$, so that $z_{\alpha/2} = z_{0.05} = 1.645$. A 90%
confidence interval for μ is calculated as follows:

$$\bar{x} \pm 1.645 \frac{s}{\sqrt{n}} = 126.9 \pm 1.645(1.416) = 126.9 \pm 2.3 \text{ or } (124.6, 129.2).$$

8.85 (a) Correct

(b) We will never know whether this particular interval covers the true mean μ.

This is a single realization of the random interval $\overline{X} \pm 1.645 \dfrac{S}{\sqrt{n}}$. In repeated

sampling, about 90% of such intervals will cover the true mean μ.

8.87 The alternative hypothesis H_1 is the assertion that is to be established; its opposite is the null hypothesis H_0.

(a) Let μ denote the population mean mileage. The hypotheses are:
$$H_0: \mu = 50, \quad H_1: \mu < 50$$

(b) Let μ denote the population mean number of pages per transmission. The hypotheses are: $H_0: \mu = 3.4, \quad H_1: \mu > 3.4$

(c) Let p denote the probability of success with the method. The hypotheses are:
$$H_0: p = .50, \quad H_1: p > .50$$

(d) Let μ denote the mean fill. The hypotheses are: $H_0: \mu = 16, \quad H_1: \mu \neq 16$

(e) Let μ denote the mean <u>percent</u> fat content. The hypotheses are:
$$H_0: p = 0.04, \quad H_1: p > 0.04$$

8.89 Since the claim is that $\mu \neq 32$, we formulate the hypotheses:
$$H_0: \mu = 32, \quad H_1: \mu \neq 32$$

With $n = 100$ and $\sigma = 10.6$, the test statistic (in standardized form) is
$$Z = \frac{\overline{X} - 32}{10.6/\sqrt{100}} = \frac{\overline{X} - 32}{1.06}.$$

(a) $\left|\overline{X} - 32\right| \geq 2.47$ is equivalent to $|Z| \geq \dfrac{2.47}{1.06} = 2.33$, or more simply $|Z| \geq 2.33$.

Since $P[|Z| \geq 2.33] = 0.02$, the level of significance is $\alpha = 0.02$.

(b) Since $P[|Z| \geq 1.96] = 0.05$, the rejection region $R: |Z| \geq 1.96$ has the level of significance $\alpha = 0.05$. Now, since $|Z| \geq 1.96$ is equivalent to

$\left|\overline{X} - 32\right| \geq 1.96(1.06) = 2.0776$, we conclude that $c = 2.0776$.

8.91 (a) Reject H_0 since the p-value is 0.014, which is less than $\alpha = 0.03$.

(b) To test the hypotheses: $H_0: \mu = 84 \quad H_1: \mu < 84$

we use the test statistic $\dfrac{\overline{X} - 84}{S/\sqrt{n}}$ (the observed value of which, from the

printout, is -2.45). Since H_1 is left-sided, the rejection region should have the form $R: Z \leq z_\alpha$. Since $z_{0.01} = -2.33$, the rejection region is $R: Z \leq -2.33$ (illustrated at the top of the next page). Since the value of the test statistic, namely -2.45, is in R, we reject H_0 at $\alpha = 0.01$.

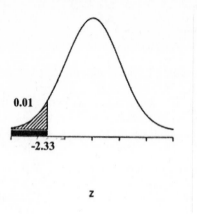

8.93 (a) $E[X] = 0(0.5) + 1(0.3) + 2(0.2) = 0.7$

(b)   Since $E[X^2] = 0^2(0.5) + 1^2(0.3) + 2^2(0.2) = 1.1$, we observe that

$\mathrm{var}[X] = E[X^2] - (E[X])^2 = 1.1 - (0.7)^2 = 0.61$, and so $\sigma = \sqrt{0.61} = 0.781$

(c)   To test $H_0: \mu = 0.7$ versus $H_1: \mu > 0.7$, we use the test statistic

$Z = \dfrac{\overline{X} - 0.7}{S/\sqrt{n}}$ with $n = 64$ and we will use the standard deviation from the

new sample because the distribution of sales may have changed. With
$\alpha = 0.05$, the rejection region is $R: Z \geq 1.645$. The observed value of the test
statistic is

$$z = \frac{0.84 - 0.7}{0.4/\sqrt{64}} = 2.80,$$

which is in R. Therefore, we reject H_0 at $\alpha = 0.05$. Furthermore, the p-value
is $P[Z \geq 2.80] = 0.0026$. So, the evidence in favor of increased mean sales is
very strong. The rejection region and p-value are illustrated below (in that
order):

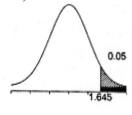

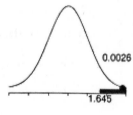

8.95 (a)   The point estimate of the proportion unemployed is $\hat{p} = \frac{175}{2000} = 0.0875$.

(b)   The 95% error margin = 1.96(Estimated S.E.), which is given by

$$1.96\sqrt{\frac{\hat{p}\hat{q}}{n}} = 1.96\sqrt{\frac{(0.0875)(0.9125)}{2000}} = 0.0124.$$

8.97 Let $\hat{p} =$ sample proportion.

(a) Test statistic: $Z = \dfrac{\hat{p} - 0.25}{\sqrt{\frac{(0.25)(0.75)}{n}}}$

Since $z_{0.05} = 1.645$ and H_1 is right-sided, the rejection region is $R: Z \geq 1.645$.

(b) The observed sample proportion is $\hat{p} = \frac{65}{190}$, and the test statistic has the value

$Z = \dfrac{0.342 - 0.25}{\sqrt{\frac{(0.25)(0.75)}{190}}} = 2.93$, which is in R. Hence, H_0 is rejected at $\alpha = 0.05$.

Furthermore, the p-value is $P[Z \geq 2.93] = 0.0017$. So, the evidence in favor of the alternative is very strong. The rejection region and p-value are illustrated below (in that order):

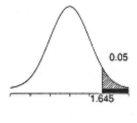
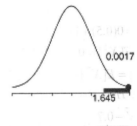

8.99 Since the intent is to establish that $p \neq 0.3$, we formulate the following hypotheses: $H_0 : p = 0.3, \quad H_1 : p \neq 0.3$

Denoting the sample proportion by $\hat{p}$, the test statistic is $Z = \dfrac{\hat{p} - 0.3}{\sqrt{\frac{(0.3)(0.7)}{250}}} = \dfrac{\hat{p} - 0.3}{0.029}$

(a) Using the test statistic $|\hat{p} - 0.3| \geq 0.06$ implies that $|Z| \geq \dfrac{0.06}{0.029} = 2.07$, or more simply, $|Z| \geq 2.07$. Since $P[|Z| \geq 2.07] = 2P[Z \leq -2.07] = 2(0.0192)$ $= 0.0384$, we have $\alpha = 0.0384$.

(b) Because $P[Z \geq 1.645] = 0.05$, the rejection region with $\alpha = 0.10$ and H_1 two-sided is $R: |Z| \geq 1.645$. Now, $|Z| \geq 1.645$ implies that

$$|\hat{p} - 0.3| \geq 1.645(0.029) = 0.048.$$

Therefore, $c = 0.048$.

8.101 (a) Let p denote the population proportion in favor of almond scent. Since the intent is to establish that $p \neq 0.5$, we formulate the following hypotheses:

$H_0 : p = 0.5, \quad H_1 : p \neq 0.5$ The test statistic is $Z = \dfrac{\hat{p} - 0.5}{\sqrt{\frac{(0.5)(0.5)}{250}}} = \dfrac{\hat{p} - 0.5}{0.0316}$.

With $\alpha = 0.05$, $z_{0.05/2} = 1.96$ and H_1 is two-sided, the rejection region is $R: |Z| \geq 1.96$. The observed sample proportion is $\hat{p} = \frac{105}{250} = 0.42$, and the test statistic has the value $Z = \dfrac{0.42 - 0.5}{0.0316} = -2.53$, which is in R. Hence, H_0 is rejected at $\alpha = 0.05$. So, we conclude that the popularity of the two scents is significantly different.

(b) A 95% confidence interval for p is given by
$$\hat{p} \pm 1.96\sqrt{\tfrac{\hat{p}\hat{q}}{n}} = 0.42 \pm 1.96\sqrt{\tfrac{(0.42)(0.58)}{250}} = 0.42 \pm 0.06 \text{ or } (0.36, 0.48).$$

8.103 (a) Let p denote the population proportion of plants that are of the dwarf variety. Since the intent is to establish that $p \neq 0.8$, we formulate the following hypotheses: $H_0 : p = 0.8$, $H_1 : p \neq 0.8$ The test statistic for a sample of size $n = 200$ is $Z = \dfrac{\hat{p} - 0.8}{\sqrt{\frac{(0.8)(0.2)}{200}}} = \dfrac{\hat{p} - 0.8}{0.0283}$. With $\alpha = 0.05$, $z_{0.05/2} = 1.96$ and H_1 is two-sided, the rejection region is $R: |Z| \geq 1.96$. The observed sample proportion is $\hat{p} = \frac{136}{200} = 0.68$, and the test statistic has the value $Z = \dfrac{0.68 - 0.8}{0.0283} = -4.24$, which is in R. Hence, H_0 is rejected at $\alpha = 0.05$. Furthermore, the associated p-value is given by
$$P[Z \leq -4.24] + P[Z \geq 4.24] < 0.0001,$$
which is extremely small. As such, a contradiction of the genetic model is strongly indicated.

(b) A 95% confidence interval for p is given by
$$\hat{p} \pm 1.96\sqrt{\tfrac{\hat{p}\hat{q}}{n}} = 0.68 \pm 1.96\sqrt{\tfrac{(0.68)(0.32)}{200}} = 0.68 \pm 0.06 \text{ or } (0.62, 0.74).$$

8.105 (a) Under the alternative $\mu_1 = 10.5$, $\overline{X}$ is normally distributed with mean 10.5 and standard deviation $= \frac{\sigma}{\sqrt{n}} = 0.25$. So, $Z = \dfrac{\overline{X} - 10.5}{0.25}$ is $N(0,1)$. Therefore, the power at $\mu_1 = 10.5$ is given by
$$P[\overline{X} \geq 10.49 \text{ when } \mu_1 = 10.5] = P[Z \geq \tfrac{10.49 - 10.5}{0.25}] = P[Z \geq -0.04] = 0.5160.$$

(b) For $\mu_1 = 10.8$, $Z = \dfrac{\overline{X} - 10.8}{0.25}$ is $N(0,1)$. Therefore, the power at $\mu_1 = 10.8$ is given by
$$P[\overline{X} \geq 10.49 \text{ when } \mu_1 = 10.8] = P[Z \geq \tfrac{10.49 - 10.8}{0.25}]$$
$$= P[Z \geq -1.24] = 0.8925.$$

8.107 (a) A 99% confidence interval will be larger than the 95% confidence interval given in the printout. In fact, the greater the confidence <u>level</u>, the larger the confidence <u>interval</u> will be for a given sample size. To verify this, observe that $\bar{x} \pm z_{0.01/2} \dfrac{s}{\sqrt{n}} = 77.458 \pm 2.58 \dfrac{1.101}{\sqrt{40}} = 77.458 \pm 0.4491$,

or (77.001, 77.907).

(b) To test $H_0 : \mu = 77$ versus $H_1 : \mu > 77$, we use the test statistic

$Z = \dfrac{\overline{X} - 77}{1.101/\sqrt{40}}$. With $\alpha = 0.01$, since H_1 is right-sided, the rejection region is $R : Z \geq 2.58$ (illustrated below). The observed value of the test statistic is

$$z = \frac{77.458 - 77}{1.101/\sqrt{40}} = 2.630,$$

which is in R. Therefore, we reject H_0 at $\alpha = 0.01$.

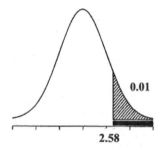

0.01

2.58

z

8.109 Enter the data into column C2 in a Minitab worksheet.

(a) Run a 1-sample t-test since the population standard deviation is unknown. The output is as follows.

T Confidence Intervals

Variable	N	Mean	StDev	SE Mean	97.0 % CI
C2	40	1.7150	0.4748	0.0751	(1.5459, 1.8841)

So, the 97% confidence interval is (1.5459, 1.8841).

(b) Testing the hypotheses $H_0 : \mu = 1.9$ versus $H_1 : \mu \neq 1.9$ using Minitab yields the following output:

T-Test of the Mean

Test of mu = 1.9000 vs mu not = 1.9000

Variable	N	Mean	StDev	SE Mean	T	P
C2	40	1.7150	0.4748	0.0751	-2.46	0.018

Since the p-value is 0.018, we reject H_0 at $\alpha = 0.03$.

8.111 Enter the data into column C4 in a Minitab worksheet. Run a 1-sample t-test since the population standard deviation is unknown. The output is as follows.

T Confidence Intervals

```
Variable     N      Mean    StDev    SE Mean      90.0 % CI
C4          28    110.39    23.18     4.38    (102.93, 117.85)
```

So, the 90% confidence interval is (102.93, 117.85).

8.113 (a) Enter the data into column C6 in a Minitab worksheet. Run a 1-sample t test since the population standard deviation is unknown. The output is as follows.

T Confidence Intervals

```
Variable     N      Mean    StDev    SE Mean      95.0 % CI
C6         151     4.072    9.065    0.738    ( 2.614,  5.530)
```

So, the 95% confidence interval is (2.614, 5.530).

(b) The histogram is as follows. Note that the histogram has a long tail to the right (much more extreme than in Exercise 8.112).

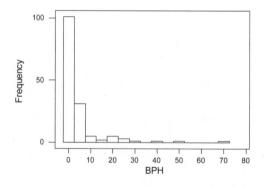

Chapter 9

SMALL-SAMPLE INFERENCES FOR NORMAL POPULATIONS

9.1   (a) $t_{0.05} = 1.943$

    (b) $-t_{0.025} = -2.110$

    (c) $-t_{0.01} = -2.764$

    (d) $t_{0.10} = 1.341$

9.3   (a) 90th percentile of $t = t_{0.10} = 1.372$

    (b) 99th percentile of $t = t_{0.01} = 3.747$

    (c) 5th percentile of $t = -t_{0.05} = -1.711$

    (d) upper quartile $= t_{0.25} = 0.688$

       lower quartile $= -t_{0.25} = -0.688$

9.5   (a) Since the area to the right of b is $1 - 0.95 = 0.05$, b is the upper 0.05 point of the t distribution with d.f. = 7. From the t-table we find that $b = t_{0.05} = 1.895$.

    (b) Since $P[T > b] = 0.025$, from the t-table with d.f. = 16, we find that
$b = t_{0.025} = 2.120$.

    (c) $b = t_{0.01} = 2.821$

    (d) Since $P[T \le b] = 0.01$, b is the lower 0.01 point of the t-distribution, namely
$b = -t_{0.01} = -2.681$.

9.7   (b) In the t-table with d.f. = 16, we look for the percentage points that are close to the given number 1.9. We find that $t_{0.05} = 1.746$ and $t_{0.025} = 2.120$. Because 1.9 lies between $t_{0.025}$ and $t_{0.05}$, the probability $P[T > 1.9]$ must lie between 0.025 and 0.05.

(c) The number 1.5 is between $t_{0.10} = 1.363$ and $t_{0.075} = 1.548$, and $P[T < -1.5] = P[T > 1.5]$. Thus, the probability $P[T < -1.5]$ must lie between 0.05 and 0.075.

(d) The number 1.9 is between $t_{0.05} = 1.812$ and $t_{0.025} = 2.228$, and $P[|T| > 1.9] = 2P[T > 1.9]$. Thus, the probability $P[|T| > 1.9]$ must lie between 2(0.025) and 2(0.05); that is, between 0.05 and 0.10.

(e) With d.f. $= 17$, the number 2.8 is between $t_{0.01} = 2.567$ and $t_{0.005} = 2.898$, so the probability $P[T > 2.8]$ must lie between 0.005 and 0.01. Because $P[|T| < 2.8] = 1 - 2P[T > 2.8]$, this probability must lie between $1 - 2(0.005) = 0.99$ and $1 - 2(0.01) = 0.98$.

9.9 (a) A 98% confidence interval for μ is given by $\overline{X} \pm t_{0.01} \dfrac{S}{\sqrt{n}}$, where d.f. $= n - 1$.

For d.f. $= 19$, we find that $t_{0.01} = 2.539$. In such case, the confidence interval is

$$140 \pm 2.539 \left(\frac{8}{\sqrt{20}} \right) = 140 \pm 4.5 \text{ or } (135.5, 144.5).$$

(b) Center is $\overline{x} = 140$. Length is $2(4.5) = 9.0$.

(c) Usually different since the length of the interval depends on the sample standard deviation S which varies from sample to sample.

9.11 Assume a normal population. A 95% confidence interval for the true mean μ is given by $\overline{X} \pm t_{0.025} \dfrac{S}{\sqrt{n}}$, where d.f. $= n - 1$. For sample size $n = 20$, we have d.f. $= n - 1 = 19$ and $t_{0.025} = 2.093$. From the sample data, we calculate $\overline{x} = 137.60$ and $s = 20.143$. The 95% confidence interval for μ is then given by

$$137.60 \pm 2.093 \left(\frac{20.143}{\sqrt{20}} \right) = 137.60 \pm 9.43 \text{ or } (128.17, 147.03).$$

9.13 Assume a normal population. A 95% confidence interval for the true mean μ is given by $\overline{X} \pm t_{0.025} \dfrac{S}{\sqrt{n}}$, where d.f. $= n - 1$. For sample size $n = 18$, we have d.f. $= n - 1 = 17$ and $t_{0.025} = 2.110$. From the sample data, we calculate $\overline{x} = 3.6$ kg and $s = 0.8$ kg. The 95% confidence interval for the mean yield μ is then given by

$$3.6 \pm 2.110 \left(\frac{0.8}{\sqrt{18}} \right) = 3.6 \pm 0.4 \text{ or } (3.2, 4.0) \text{ kg.}$$

9.15 Assume a normal population. A 95% confidence interval for the true mean μ is given by $\overline{X} \pm t_{0.025} \dfrac{S}{\sqrt{n}}$, where d.f. = $n-1$. In this case, $n = 12$, so that d.f. = $n-1 = 11$ and $t_{0.025} = 2.201$.

(a) A calculation of $\overline{x} \pm 2.201 \dfrac{s}{\sqrt{12}}$ has yielded the result (18.6, 26.2). This interval has

$$\text{center} = \overline{x} = \frac{18.6 + 26.2}{2} = 22.4$$

$$\text{half-width} = 2.201 \frac{s}{\sqrt{12}} = 22.4 - 18.6 = 3.8.$$

From the last relation, we see that $s = 3.8 \left(\dfrac{\sqrt{12}}{2.201} \right) = 5.98$.

(b) A 98% confidence interval for the true mean μ is given by $\overline{X} \pm t_{0.01} \dfrac{S}{\sqrt{n}}$, where d.f. = $n-1$. For sample size $n = 12$, we have d.f. = $n-1 = 11$ and $t_{0.01} = 2.718$. From part (a), we have $\overline{x} = 22.4$ and $s = 5.98$. The 98% confidence interval for μ is then given by

$$22.4 \pm 2.718 \left(\frac{5.98}{\sqrt{12}} \right) = 22.4 \pm 4.7 \quad \text{or} \quad (17.7, 27.1).$$

9.17 Assume a normal population. A 99% confidence interval for the true mean μ is given by $\overline{X} \pm t_{0.005} \dfrac{S}{\sqrt{n}}$, where d.f. = $n-1$. For sample size $n = 23$, we have d.f. = $n-1 = 22$ and $t_{0.005} = 2.819$. From the sample data, we have $\overline{x} = 5.483$ and $s = 0.1904$. So, the 99% confidence interval for μ is then given by

$$5.483 \pm 2.819 \left(\frac{0.1904}{\sqrt{23}} \right) = 5.483 \pm 0.111 \quad \text{or} \quad (5.372, 5.594).$$

9.19 We are to test the hypotheses $H_0 : \mu = 620$ versus $H_1 : \mu > 620$ with $\alpha = 0.05$. The test statistic is $T = \dfrac{\overline{X} - 620}{S / \sqrt{10}}$. For sample size $n = 10$, we have d.f. = $n-1 = 9$ and $t_{0.05} = 1.833$. Since H_1 is right-sided, the rejection region is $R : T \geq 1.833$. From the sample data, we have $\overline{x} = 679.0$ and $s = 98.342$. The value of the observed t is then $t = \dfrac{679.0 - 620}{98.342 / \sqrt{10}} = 1.90$, which lies in R. Hence, H_0 is rejected at $\alpha = 0.05$. Furthermore, the associated p-value is $P[T \geq 1.90] = 0.0449$. The rejection region and p-value are illustrated on the next page (in that order):

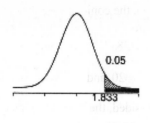
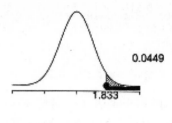

9.21 (a) Uncertain since the true mean length (a <u>population</u> parameter) is not known.
Refer to the text pages 301 – 302 for a detailed explanation.

   (b) The very definition of a confidence interval ensures this statement is true – see
property 2 on page 302 of the text.

9.23 (a) Assume that the mound diameters are normally distributed. Here, the
conjecture is that μ is larger than 21 feet, so we formulate the hypotheses:

$$H_0 : \mu = 21 \text{ versus } H_1 : \mu > 21$$

The test statistic is $T = \dfrac{\overline{X} - 21}{S/\sqrt{13}}$. For sample size $n = 13$ and $\alpha = 0.01$, we have

d.f. $= n - 1 = 12$ and $t_{0.01} = 2.681$. Since H_1 is right-sided, the rejection region is
$R : T \geq 2.681$. From the sample data, we have $\overline{x} = 26.62$ and $s = 6.56$. The

value of the observed t is then $t = \dfrac{26.62 - 21}{6.56/\sqrt{13}} = 3.09$, which lies in R. Hence, H_0

is rejected at $\alpha = 0.01$. Furthermore, the associated p-value is
$P[T \geq 3.09] = 0.0047$, so there is strong evidence in support of the conjecture.
The rejection region and p-value are illustrated below (in that order):

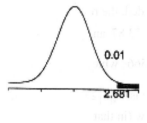
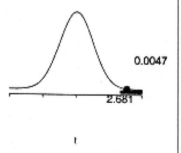

   (b) A 90% confidence interval for the true mean μ is given by $\overline{X} \pm t_{0.05} \dfrac{S}{\sqrt{n}}$,

where d.f. $= n - 1$. For sample size $n = 13$, we have d.f. $= n - 1 = 12$ and
$t_{0.05} = 1.782$. From the sample data, we have $\overline{x} = 26.62$ feet and $s = 6.56$ feet.
So, the 90% confidence interval for μ is then given by

$$26.62 \pm 1.782 \left(\frac{6.56}{\sqrt{13}} \right) = 26.62 \pm 3.24 \text{ or } (23.38, 29.86) \text{ feet.}$$

9.25 Assume that the data are normally distributed. Here, the conjecture is that μ is larger than 128 mm, so we formulate the hypotheses:
$$H_0 : \mu = 128 \text{ versus } H_1 : \mu > 128$$
The test statistic is $T = \dfrac{\overline{X} - 128}{S / \sqrt{20}}$. For sample size $n = 20$ and $\alpha = 0.05$, we have d.f. $= n - 1 = 19$ and $t_{0.05} = 1.729$. Since H_1 is right-sided, the rejection region is $R : T \geq 1.729$. From the sample data, we have $\overline{x} = 137.60$ and $s = 20.143$. The value of the observed t is then $t = \dfrac{137.60 - 128}{20.143 / \sqrt{20}} = 2.13$, which does not lie in R. Hence, H_0 is not rejected at $\alpha = 0.05$. Furthermore, the associated p-value is $P[T \geq 2.13] = 0.023$, there is still strong evidence in support of the conjecture. The rejection region and p-value are illustrated below (in that order):

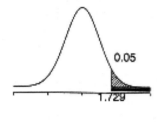
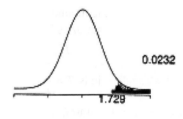

9.27 Assume a normal population. We are to test the hypotheses:
$$H_0 : \mu = 83 \text{ versus } H_1 : \mu \neq 83$$
The test statistic is $T = \dfrac{\overline{X} - 83}{S / \sqrt{8}}$. For sample size $n = 8$ and $\alpha = 0.05$, we have d.f. $= n - 1 = 7$ and $t_{0.025} = 2.365$. Since H_1 is two-sided, the rejection region is $R : |T| \geq 2.365$. From the sample data, we have $\overline{x} = 73.87$ and $s = 10.063$. The value of the observed t is then $t = \dfrac{73.87 - 83}{10.063 / \sqrt{8}} = -2.566$, which lies in R. Hence, H_0 is rejected at $\alpha = 0.05$. Also, the associated p-value is $P[|T| \geq 2.566] = 0.0372$. The rejection region and p-value are illustrated below (in that order):

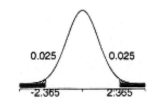
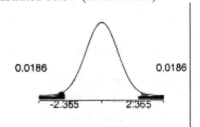

9.29 (a) Assume that the mean drying time data are normally distributed. Here, we wish to establish a decrease in the mean drying time, so we formulate the hypotheses: $H_0 : \mu = 90$ versus $H_1 : \mu < 90$

The test statistic is $T = \dfrac{\overline{X} - 90}{S/\sqrt{15}}$. For sample size $n = 15$ and $\alpha = 0.05$, we have d.f. $= n - 1 = 14$ and $-t_{0.05} = -1.761$. Since H_1 is left-sided, the rejection region is $R : T \leq -1.761$. From the sample data, we have $\overline{x} = 86$ and $s = 4.5$. The value of the observed t is then $t = \dfrac{86 - 90}{4.5/\sqrt{15}} = -3.44$, which lies in R. Hence, H_0 is rejected at $\alpha = 0.05$. Furthermore, scanning the t-table for d.f. = 14, we find that the claim would also be rejected at $\alpha = 0.005$ since $t_{0.005} = 2.977$. In fact, the p-value is approximately 0.002. This extremely small p-value lends a strong support for H_1. The rejection region and p-value are illustrated below (in that order):

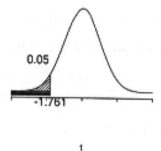
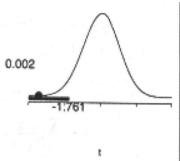

(b) A 98% confidence interval for the mean drying time μ is given by $\overline{X} \pm t_{0.01} \dfrac{S}{\sqrt{n}}$, where d.f. $= n - 1$. For sample size $n = 15$, we have d.f. $= n - 1 = 14$ and $t_{0.01} = 2.624$. From the sample data, we have $\overline{x} = 86$ and $s = 4.5$. So, the 98% confidence interval for μ is then given by $86 \pm 2.624 \left(\dfrac{4.5}{\sqrt{15}} \right) = 86 \pm 3$ or (83, 89).

9.31 (a) Reject H_0 since the p-value is 0.022, which is less than $\alpha = 0.05$.
   (b) We could have made a Type I error (if H_0 happened to be true).
   (c) Prior to sampling, the probability of making a Type I error is 0.05.
   (d) If we took more and more samples and ran the same test of hypothesis, about 5% of the time we would make a Type I error (i.e., reject a true H_0).

9.33 (a) The value $\mu_0 = 81$ is outside the 90% confidence interval (67.13, 80.62). Therefore, the null hypothesis $H_0 : \mu_0 = 81$ is rejected at $\alpha = 0.10$.
   (b) The value $\mu_0 = 69$ lies inside the 90% confidence interval (67.13, 80.62). Therefore, the null hypothesis $H_0 : \mu_0 = 69$ is not rejected at $\alpha = 0.10$.

9.35 Assume a normal population.

(a) A 95% confidence interval for μ is given by $\overline{X} \pm t_{0.025}\dfrac{S}{\sqrt{n}}$, where d.f. $= n-1$.

For sample size $n = 8$, we have d.f. $= n-1 = 7$ and $t_{0.025} = 2.365$. From the sample data, we have $\overline{x} = 6.78$ and $s = 6.58$. So, the 95% confidence interval for μ is then given by

$$6.78 \pm 2.365\left(\frac{6.58}{\sqrt{8}}\right) = 6.78 \pm 5.50 \quad \text{or} \quad (1.28,\ 12.28).$$

(b) The value $\mu_0 = 15$ is outside the 95% confidence interval. Therefore, the null hypothesis $H_0 : \mu_0 = 15$ is rejected at $\alpha = 0.05$.

(c) Here, we wish to test the hypotheses: $H_0 : \mu = 15$ versus $H_1 : \mu \neq 15$

The test statistic is $T = \dfrac{\overline{X} - 15}{S/\sqrt{8}}$. For sample size $n = 8$ and $\alpha = 0.05$, we have

d.f. $= n-1 = 7$ and $t_{0.025} = 2.365$. Since H_1 is two-sided, the rejection region is $R : |T| \geq 2.365$. From the sample data, we have $\overline{x} = 6.78$ and $s = 6.58$. The

value of the observed t is then $t = \dfrac{6.78 - 15}{6.58/\sqrt{8}} = -3.53$, which lies in R. Hence, H_0

is rejected at $\alpha = 0.05$. This confirms the conclusion in part (b).

9.37 The acceptance region of the $\alpha = 0.05$ test is $-1.96 \leq \dfrac{\overline{X} - \mu_0}{S/\sqrt{n}} \leq 1.96$. A

rearrangement of these inequalities subsequently yields

$$\overline{X} - 1.96\frac{S}{\sqrt{n}} \leq \mu_0 \leq \overline{X} + 1.96\frac{S}{\sqrt{n}}.$$

Thus, any μ_0 that lies in the interval $\overline{X} \pm 1.96\dfrac{S}{\sqrt{n}}$ will not be rejected at $\alpha = 0.05$.

This interval is precisely the 95% confidence interval for μ.

9.39 (a) Since the area to its left is $1 - 0.10 = 0.90$, it is the 90th percentile $\chi^2_{0.10} = 23.54$.

(b) The 5th percentile is $\chi^2_{0.95} = 12.34$.

(c) $\chi^2_{0.10} = 40.26$

(d) $\chi^2_{0.95} = 3.33$

9.41 (a) The sample standard deviation $S = 3.3615$ estimates σ.

(b) For d.f. $= n - 1 = 4$, we have $\chi^2_{0.025} = 11.14$ and $\chi^2_{0.975} = 0.48$. The general form

for a 95% confidence interval for σ is given by $\left(S\sqrt{\dfrac{n-1}{\chi^2_{\%}}} \, , \, S\sqrt{\dfrac{n-1}{\chi^2_{1-\%}}} \right)$. In the

present problem, the confidence interval is

$$\left(3.3615\sqrt{\frac{4}{11.14}} \, , \, 3.3615\sqrt{\frac{4}{0.48}} \right) = (2.01,\ 9.70).$$

(c) The center of the confidence interval in part (b) is $\frac{2.01 + 9.70}{2} = 5.855$, which is not

the same as $s = 3.3615$.

9.43 The population distribution is assumed to be normal. We wish to test the
hypotheses: $H_0 : \sigma = 0.6$ versus $H_1 : \sigma < 0.6$. The test statistic is

$$\chi^2 = \frac{(n-1)S^2}{\sigma^2} = \frac{(n-1)S^2}{(0.6)^2}.$$ For sample size $n = 40$ and $\alpha = 0.05$, we have

d.f. $= n - 1 = 39$ and $\chi^2_{0.95} \approx 25.7$. Since H_1 is left-sided, the rejection region is

$R : \chi^2 \leq \chi^2_{0.95} \approx 25.7$. From the sample data, we have $s = 0.475$. The value of the

observed χ^2 is then $\chi^2 = \dfrac{39(0.475)^2}{(0.6)^2} = 24.4$, which lies in R. Hence, H_0 is rejected

at $\alpha = 0.05$. As such, there is strong evidence that the red pine population standard
deviation is smaller than 0.6.

9.45 The population distribution is assumed to be normal. A 95% confidence interval

for σ is given by $\left(S\sqrt{\dfrac{n-1}{\chi^2_{0.025}}} \, , \, S\sqrt{\dfrac{n-1}{\chi^2_{0.975}}} \right)$, where d.f. $= n - 1$. For sample size

$n = 10$, we have d.f. $= n - 1 = 9$ and $\chi^2_{0.025} = 19.02$ and $\chi^2_{0.975} = 2.70$. From the
sample data, we have $s = 2.2706$. So, the 90% confidence interval for σ is given
by

$$\left(S\sqrt{\frac{n-1}{\chi^2_{0.025}}} \, , \, S\sqrt{\frac{n-1}{\chi^2_{0.975}}} \right) = \left(2.2706\sqrt{\frac{9}{19.02}} \, , \, 2.2706\sqrt{\frac{9}{2.70}} \right) = (1.56,\ 4.15).$$

9.47 (a)   A 90% confidence interval for σ is given by $\left(S\sqrt{\dfrac{n-1}{\chi^2_{0.05}}} \, , \, S\sqrt{\dfrac{n-1}{\chi^2_{0.95}}} \right)$, where

d.f. $= n - 1$. For the lizard length data, we have sample size $n = 20$, so that
d.f. $= n - 1 = 19$ and $\chi^2_{0.05} = 30.14$ and $\chi^2_{0.95} = 10.12$. From the sample data,
we have $s = 20.143$. So, the 90% confidence interval for σ is given by

$$\left(S\sqrt{\frac{n-1}{\chi^2_{0.05}}} \; , \; S\sqrt{\frac{n-1}{\chi^2_{0.95}}} \right) = \left(20.143\sqrt{\frac{19}{30.14}} \; , \; 20.143\sqrt{\frac{19}{10.12}} \right) = (15.99, \, 27.60).$$

(b)   The value $\sigma_0 = 9$ is not in the 90% confidence interval, so the null

hypothesis $H_0 : \sigma = 9$ is rejected at $\alpha = 0.10$.

9.49 A 95% confidence interval for σ is given by $\left(S\sqrt{\dfrac{n-1}{\chi^2_{0.025}}} \; , \; S\sqrt{\dfrac{n-1}{\chi^2_{0.975}}} \right)$, where

d.f. $= n-1$. For this data, we have sample size $n = 13$, so that d.f. $= n-1 = 12$ and
$\chi^2_{0.025} = 23.34$ and $\chi^2_{0.975} = 4.40$. From the sample data, we have $s = 6.56$. So, the
95% confidence interval for σ is given by

$$\left(S\sqrt{\frac{n-1}{\chi^2_{0.025}}} \; , \; S\sqrt{\frac{n-1}{\chi^2_{0.975}}} \right) = \left(6.56\sqrt{\frac{12}{23.34}} \; , \; 6.56\sqrt{\frac{12}{4.40}} \right) = (4.70, \, 10.83).$$

9.51 (a) $t_{0.05} = 2.353$

(b) $t_{0.025} = 2.179$

(c) $-t_{0.05} = -2.353$

(d) $-t_{0.05} = -1.782$

9.53 (a) From the t-table we find that, with d.f. = 12, $t_{0.05} = 1.782$. Now, using the
normal table, we interpolate to obtain
$$P[Z > t_{0.05}] = P[Z > 1.782] = 0.037.$$
Note that this probability is smaller than $P[T > t_{0.05}] = 0.05$.

(b) For d.f. = 5, we have $t_{0.05} = 2.015$. Using the normal table, we find that
$P[Z > 2.015] = 0.022$, which is smaller than 0.05. Next, for d.f. = 20, we have
$t_{0.05} = 1.725$, while the normal table gives $P[Z > 1.725] = 0.042$. We observe
that the probability $P[Z > t_{0.05}]$ is always less than 0.05, but the difference
decreases as the d.f. increases.

9.55 Assume that the measurements are normally distributed. A 99% confidence
interval for the true mean μ is given by $\overline{X} \pm t_{0.005} \dfrac{S}{\sqrt{n}}$, where d.f. $= n-1$. For
sample size $n = 14$, we have d.f. $= n-1 = 13$ and $t_{0.005} = 3.012$. From the sample
data, we calculate $\overline{x} = 47$ and $s = 9.4$. The 99% confidence interval for the mean
measurement μ is then given by

$$47 \pm 3.012 \left(\frac{9.4}{\sqrt{14}} \right) = 47 \pm 7.6 \ \text{ or } \ (39.4, \, 54.6).$$

9.57 Assume a normal population. A 99% confidence interval for the true mean μ is given by $\overline{X} \pm t_{0.005} \dfrac{S}{\sqrt{n}}$, where d.f. $= n-1$. In this case, $n = 11$, so that d.f. $= n-1 = 10$ and $t_{0.005} = 3.169$.

(a) A calculation of $\overline{x} \pm 3.169 \dfrac{s}{\sqrt{11}}$ has yielded the result (62.5, 86.9). This interval has

$$\text{center} = \overline{x} = \frac{62.5 + 86.9}{2} = 74.7$$

$$\text{half-width} = 3.169 \frac{s}{\sqrt{11}} = 86.9 - 74.7 = 12.2 .$$

From the last relation, we see that $\dfrac{s}{\sqrt{11}} = \dfrac{12.2}{3.169}$. Furthermore, a point estimate of μ is $\overline{X} = 74.7$, and the 95% error margin is $t_{0.025} \dfrac{S}{\sqrt{11}} = 2.228 \left(\dfrac{12.2}{3.169} \right) = 8.58$.

(b) A 90% confidence interval for the true mean μ is given by $\overline{X} \pm t_{0.05} \dfrac{S}{\sqrt{n}}$, where d.f. $= n-1$. For sample size $n = 11$, we have d.f. $= n-1 = 10$ and $t_{0.05} = 1.812$. From part (a), we have $\overline{x} = 74.7$ and $\dfrac{s}{\sqrt{11}} = \dfrac{12.2}{3.169}$. The 90% confidence interval for μ is then given by

$$74.7 \pm 1.812 \left(\frac{12.2}{3.169} \right) = 74.7 \pm 7.0 \ \text{ or } \ (67.7, 81.7) .$$

9.59 Assume a normal population. We are to test the hypotheses
$H_0 : \mu = 42$ versus $H_1 : \mu < 42$ with $\alpha = 0.01$. The test statistic is $T = \dfrac{\overline{X} - 42}{S/\sqrt{21}}$.
For sample size $n = 21$, we have d.f. $= n-1 = 20$ and $-t_{0.01} = -2.528$. Since H_1 is left-sided, the rejection region is $R : T \le -2.528$. From the sample data, we have $\overline{x} = 38.4$ and $s = 5.1$. The value of the observed t is then $t = \dfrac{38.4 - 42}{5.1/\sqrt{21}} = -3.23$,
which lies in R. Hence, H_0 is rejected at $\alpha = 0.01$. Furthermore, the associated p-value is $P[T \le -3.23] < 0.005$. So, there is strong evidence that the mean time to blossom is less than 42 days. The rejection region and p-value are illustrated on the top of the next page (in that order):

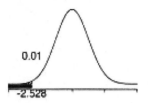 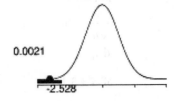

9.61 Since H_1 is two-sided, we need a two-sided rejection region. For sample size $n = 18$ and $\alpha = 0.02$, we have d.f. $= n - 1 = 17$ and $t_{0.01} = 2.567$. So, the rejection region in this case is $R : |T| \geq 2.567$ (which is illustrated below). Since the observed value of t, namely 1.59, does not lie in R, H_0 is not rejected at $\alpha = 0.02$.

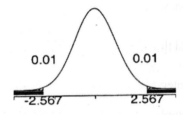

9.63 (a) Assume a normal population. Let μ denote the mean potency after exposure. The supplier's claim is valid if $\mu > 65$, and this is to be demonstrated. So, we are to test the hypotheses $H_0 : \mu = 65$ versus $H_1 : \mu > 65$ with $\alpha = 0.05$. The test statistic is $T = \dfrac{\overline{X} - 65}{S / \sqrt{9}}$. For sample size $n = 9$, we have d.f. $= n - 1 = 8$ and $t_{0.05} = 1.860$. Since H_1 is right-sided, the rejection region is $R : T \geq 1.860$ (which is illustrated below).

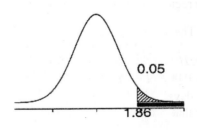

(b)   From the sample data, we have $\bar{x} = 65.22$ and $s = 3.67$. The value of the
observed t is then $t = \dfrac{65.22 - 65}{3.67 / \sqrt{9}} = 0.18$, which does not lie in R. Hence, H_0 is
not rejected at $\alpha = 0.05$. In fact, a comparison of the rejection region (left
boundary of 1.860) and this value of the t-statistic (0.18) clearly shows that
the null hypotheses would not be rejected at any reasonable level of α. As
such, the claim that $\mu > 65$ is not demonstrated.

9.65   Assume a normal population. We want to establish that $\mu > 1500$ (i.e., the
advertiser's claim is false). So, we are to test the hypotheses
$H_0 : \mu = 1500$ versus $H_1 : \mu > 1500$ with $\alpha = 0.05$. The test statistic is
$T = \dfrac{\bar{X} - 1500}{S / \sqrt{5}}$. For sample size $n = 5$, we have d.f. $= n - 1 = 4$ and $t_{0.05} = 2.132$.
Since H_1 is right-sided, the rejection region is $R : T \geq 2.132$. From the sample
data, we have $\bar{x} = 1620$ and $s = 90$. The value of the observed t is then
$t = \dfrac{1620 - 1500}{90 / \sqrt{5}} = 2.98$, which lies in R. Hence, H_0 is rejected at $\alpha = 0.05$.
Furthermore, scanning the t-table for d.f. = 4, we find that 2.98 lies between
$t_{0.025} = 2.776$ and $t_{0.01} = 3.747$. So, the associated p-value is $P[T \geq 2.98]$ is
between 0.01 and 0.025, near about 0.02. As such, we conclude that the
advertiser's claim is strongly contradicted. The rejection region and p-value are
illustrated below (in that order):

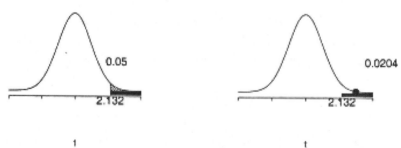

9.67   We use Minitab for this exercise. One could also complete it by hand by
mimicking the approach of Exercise 9.27, for instance.
(a)   Enter the sample data into column C1 of a Minitab worksheet. The output
is as follows:

T Confidence Intervals

Variable	N	Mean	StDev	SE Mean	95.0 % CI
C1	7	439.57	17.82	6.74	(423.08, 456.06)

So, the 95% confidence interval is (423.08, 456.06).

(b) In order to test the hypotheses $H_0 : \mu = 453$ versus $H_1 : \mu \neq 453$, run a one-sample t test on Minitab (using the data in column C1) to obtain:

T-Test of the Mean
```
Test of mu = 453.00 vs mu not = 453.00
Variable     N     Mean    StDev    SE Mean     T       P
C1           7    439.57   17.82     6.74     -1.99   0.093
```

Hence, since the p-value is 0.093, we (barely) reject $_0$ at $\alpha = 0.10$.

9.69 (a) $\chi^2_{0.05} = 12.59$

(b) $\chi^2_{0.025} = 38.08$

(c) $\chi^2_{0.95} = 1.64$

(d) $\chi^2_{0.975} = 11.69$

9.71 The population distribution is assumed to be normal. In each case, the test statistic is $\chi^2 = \dfrac{(n-1)S^2}{\sigma^2} = \dfrac{(n-1)S^2}{(1.0)^2}$, with d.f. $= n-1$, and the rejection region has the form $R : \chi^2 \geq \chi^2_{0.05}$.

(a) For sample size $n = 25$ and $\alpha = 0.05$, we have d.f. $= n - 1 = 24$ and $\chi^2_{0.05} = 36.42$. Observe that the value of the test statistic is

$$\chi^2 = \frac{(n-1)s^2}{(1.0)^2} = \frac{\sum(x_i - \bar{x})^2}{(1.0)^2} = \frac{40.16}{1.0} = 40.16,$$

which is greater than 36.42, and hence lies in R. Hence, H_0 is rejected at $\alpha = 0.05$.

(b) For sample size $n = 15$ and $\alpha = 0.05$, we have d.f. $= n - 1 = 14$ and $\chi^2_{0.05} = 23.68$. Observe that the value of the test statistic is

$$\chi^2 = \frac{(n-1)s^2}{(1.0)^2} = \frac{14(1.2)^2}{(1.0)^2} = 20.16,$$

which is less than 23.68, and hence does not lie in R. Hence, H_0 is not rejected at $\alpha = 0.05$.

(c) For sample size $n = 6$ and $\alpha = 0.05$, we have d.f. $= n - 1 = 5$ and $\chi^2_{0.05} = 11.07$. Observe that the value of the test statistic is

$$\chi^2 = \frac{(n-1)s^2}{(1.0)^2} = \frac{5(2.229)^2}{(1.0)^2} = 24.84,$$

which is much greater than 11.07, and hence lies in R. Hence, H_0 is rejected at $\alpha = 0.05$.

9.73 Assume a normal population. A 95% confidence interval for the true mean μ is

given by $\overline{X} \pm t_{0.025} \dfrac{S}{\sqrt{n}}$, where d.f. $= n-1$. In this case, $n = 10$, so that

d.f. $= n - 1 = 9$ and $t_{0.025} = 2.262$.

(a) A calculation of $\overline{x} \pm 2.262 \dfrac{s}{\sqrt{10}}$ has yielded the result (36.2, 45.8). This interval

has

center $= \overline{x} = \dfrac{36.2 + 45.8}{2} = 41.0$ half-width $= 2.262 \dfrac{s}{\sqrt{10}} = 45.8 - 41.0 = 4.8$.

From the last relation, we see that $s = 4.8 \left(\dfrac{\sqrt{10}}{2.262} \right) = 6.710$.

(b) A 98% confidence interval for the true mean μ is given by $\overline{X} \pm t_{0.01} \dfrac{S}{\sqrt{n}}$, where

d.f. $= n - 1$. For sample size $n = 10$, we have d.f. $= n - 1 = 9$ and $t_{0.01} = 2.821$.
From part (a), we have $\overline{x} = 41.0$ and $s = 6.710$. The 98% confidence interval

for μ is then given by $41.0 \pm 2.821 \left(\dfrac{6.710}{\sqrt{10}} \right) = 41.0 \pm 6.0$ or (35.0, 47.0).

(c) The population distribution is assumed to be normal. A 95% confidence

interval for σ is given by $\left(S \sqrt{\dfrac{n-1}{\chi^2_{0.025}}} \,,\, S \sqrt{\dfrac{n-1}{\chi^2_{0.975}}} \right)$, where d.f. $= n - 1$. For

sample size $n = 10$, we have d.f. $= n - 1 = 9$ and $\chi^2_{0.025} = 19.02$ and $\chi^2_{0.975} = 2.70$.
From part (a), we have $s = 6.710$. So, the 95% confidence interval for σ is
given by

$$\left(S \sqrt{\dfrac{n-1}{\chi^2_{0.025}}} \,,\, S \sqrt{\dfrac{n-1}{\chi^2_{0.975}}} \right) = \left(6.710 \sqrt{\dfrac{9}{19.02}} \,,\, 6.710 \sqrt{\dfrac{9}{2.70}} \right) = (4.62, 12.25).$$

9.75 (a) Uncertain since the true mean length (a <u>population</u> parameter) is not known.
        Refer to the text pages 301 – 302 for a detailed explanation.
     (b) The very definition of a confidence interval ensures this statement is true – see
        property 2 on page 302 of the text.

9.77 Use only those values that are coded with a 1 (for Male). Enter into column C2 of a
     Minitab worksheet – here is the output.

T-Test of the Mean
Test of mu = 2.700 vs mu > 2.700

Variable	N	Mean	StDev	SE Mean	T	P
C2	20	2.964	0.525	0.117	2.25	0.018

Hence, we reject H_0 at $\alpha = 0.025$ since the p-value is less than 0.025.

Chapter 10
COMPARING TWO TREATMENTS

10.1 First group using first letter:

    {B, C}   {B, E}   {B, H}   {B, P}
    {C, E}   {C, H}   {C, P}
    {E, H}   {E, P}
    {H, P}

10.3 (a) We use the first letter of the first names:

    { (S, G), (T, E) }   { (S, G), (T, R) }   { (S, G), (E, R) }
    { (S, J), (T, E) }   { (S, J), (T, R) }   { (S, J), (E, R) }
    { (J, G), (T, E) }   { (J, G), (T, R) }   { (J, G), (E, R) }

    (b) There are three sets each consisting of three pairs.

10.5 (a) A point estimate of $\mu_1 - \mu_2$ is given by $\bar{x} - \bar{y} = 73 - 66 = 7$.

$$\text{Estimated S.E.} = \sqrt{\frac{s_1^2}{n_1} + \frac{s_2^2}{n_2}} = \sqrt{\frac{151}{52} + \frac{142}{44}} = 2.48$$

    (b) A large sample 95% confidence interval for $\mu_1 - \mu_2$ is given by

$$\left(\bar{x} - \bar{y}\right) \pm z_{0.05/2}\sqrt{\frac{s_1^2}{n_1} + \frac{s_2^2}{n_2}} = 7 \pm 1.96(2.48) = 7 \pm 4.9 \ \text{ or } (2.1, 11.9).$$

10.7 We are to test the hypotheses $H_0 : \mu_1 - \mu_2 = 0$ versus $H_1 : \mu_1 - \mu_2 \neq 0$ with $\alpha = 0.05$. Since the sample sizes are large, we employ the Z-test, and so, the test statistic is $Z = \dfrac{\bar{X} - \bar{Y}}{\sqrt{\dfrac{S_1^2}{n_1} + \dfrac{S_2^2}{n_2}}}$. Since H_1 is two-sided, the rejection region is

$R : |Z| \geq z_{0.05/2} = 1.96$. From the sample data, we calculate the value of the observed z

to be $z = \dfrac{76.4 - 81.2}{\sqrt{\dfrac{(8.2)^2}{90} + \dfrac{(7.6)^2}{100}}} = -\dfrac{4.8}{1.151} = -4.17$, which lies in R. Hence, H_0 is

rejected at $\alpha = 0.05$. Furthermore, the associated p-value is $P[|Z| \geq 4.17] < 0.0001$,
so the evidence against *no mean difference* is very strong.

10.9 (a) Since the assertion is that $\mu_1 < \mu_2$, we formulate the hypotheses

$H_0 : \mu_1 - \mu_2 = 0$ versus $H_1 : \mu_1 - \mu_2 < 0$.

(b) Since the sample sizes $n_1 = 78$ and $n_2 = 62$ are large, we employ the Z-test, and

so, the test statistic is $Z = \dfrac{\overline{X} - \overline{Y}}{\sqrt{\dfrac{S_1^2}{n_1} + \dfrac{S_2^2}{n_2}}}$. Since H_1 is left-sided and $\alpha = 0.05$, the

rejection region is $R : Z \leq z_{0.05} = -1.645$.

(c) From the sample data, we calculate the value of the observed z to be

$z = \dfrac{92 - 118}{\sqrt{\dfrac{(46.2)^2}{78} + \dfrac{(53.4)^2}{62}}} = -\dfrac{26}{8.565} = -3.04$, which lies in R. Hence, H_0 is

rejected at $\alpha = 0.05$. Furthermore, the associated p-value is
$P[Z \leq -3.04] < 0.0012$, so the evidence is support of H_1 is very strong.

10.11   The problem here is to test $H_0 : \mu_1 - \mu_2 = 0$ versus $H_1 : \mu_1 - \mu_2 \neq 0$ with
$\alpha = 0.05$. Because 0 is not included in the 95% confidence interval (655, 1165)
(from Exercise 10.10), the conclusion is that H_0 is rejected at $\alpha = 0.05$.

10.13   The summary statistics are:

Aggressive:	$n_1 = 47$,	$\bar{x} = 7.92$,	$s_1 = 3.45$
Non-aggressive:	$n_2 = 38$,	$\bar{y} = 5.80$,	$s_2 = 2.87$

Denote by μ_1 and μ_2 the population BPC scores for the two groups
"aggressive" and "non-aggressive", respectively. Since the conjecture is that
μ_1 is larger than μ_2, we formulate the hypotheses

$H_0 : \mu_1 - \mu_2 = 0$ versus $H_1 : \mu_1 - \mu_2 > 0$.

Since the sample sizes are large, we employ the Z-test, and so, the test statistic

is $Z = \dfrac{\overline{X} - \overline{Y}}{\sqrt{\dfrac{S_1^2}{n_1} + \dfrac{S_2^2}{n_2}}}$. Since H_1 is right-sided, the rejection region is of the form

$R : Z \geq c$. From the sample data, we calculate the value of the observed z to be

$$z = \frac{7.92 - 5.80}{\sqrt{\dfrac{(3.45)^2}{47} + \dfrac{(2.87)^2}{38}}} = \frac{2.12}{0.06856} = 3.09.$$ The p-value $P[Z \geq 3.09]$ is smaller

than 0.001. This extremely small p-value provides a strong justification of the conjecture that μ_1 is larger than μ_2.

10.15   (a)   We first obtain:

$$\bar{x} = 7 \qquad\qquad\qquad\qquad \bar{y} = 5$$

$$s_1^2 = \frac{\left(2^2 + (-2)^2 + 0^2\right)}{2} = 4 \qquad\qquad s_2^2 = \frac{\left(1^2 + (-2)^2 + 1^2\right)}{2} = 3$$

Consequently, the pooled variance is given by

$$s_{pooled}^2 = \frac{\sum (x_i - \bar{x})^2 + \sum (y_i - \bar{y})^2}{n_1 + n_2 - 2} = \frac{8 + 6}{3 + 3 - 2} = 3.5.$$

(b)   We estimate the common sigma by $s_{pooled} = \sqrt{3.5} = 1.87$.

(c)   The t-statistic is $T = \dfrac{\bar{X} - \bar{Y}}{s_{pooled}\sqrt{\dfrac{1}{n_1} + \dfrac{1}{n_2}}}$ with d.f. = $n_1 + n_2 - 2$. In the

present problem, we have $t = \dfrac{7 - 5}{1.87\sqrt{\dfrac{1}{3} + \dfrac{1}{3}}} = 1.31$, with d.f. = 4.

10.17   (a)   We are to test the hypotheses $H_0 : \mu_1 - \mu_2 = 0$ versus $H_1 : \mu_1 - \mu_2 \neq 0$ with $\alpha = 0.05$. We assume normal populations with equal variance, so the test statistic is

$$T = \frac{\bar{X} - \bar{Y}}{s_{pooled}\sqrt{\dfrac{1}{n_1} + \dfrac{1}{n_2}}}$$ with d.f. = $n_1 + n_2 - 2$

Since H_1 is two-sided, the rejection region is $R : |T| \geq t_{0.05/2} = 2.110$ (for d.f. $= 8 + 11 - 2 = 17$). From the sample data, we calculate the following:

$$\bar{x} - \bar{y} = 73.9 - 91.9 = -18, \; s_1^2 = 101.268, \text{ and } s_2^2 = 153.291$$

Consequently, we have

$$s_{pooled} = \sqrt{\frac{(n_1 - 1)s_1^2 + (n_2 - 1)s_2^2}{n_1 + n_2 - 2}} = \sqrt{\frac{(8 - 1)101.268 + (11 - 1)153.291}{8 + 11 - 2}} = 11.483.$$

As such, the value of the observed t is

$$t = \frac{73.9 - 91.9}{11.483\sqrt{\dfrac{1}{8} + \dfrac{1}{11}}} = -3.37,$$

which lies in R. So, H_0 is rejected at $\alpha = 0.05$. Furthermore, the associated p-value is $P[|T| \geq 3.37] = 0.0036$. The rejection region and p-value are illustrated below (in that order):

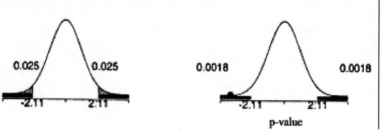

b)   A corresponding 95% confidence interval has the form

$$\left(\overline{X} - \overline{Y}\right) \pm t_{\alpha/2}\, s_{pooled} \sqrt{\frac{1}{n_1} + \frac{1}{n_2}}\ .$$

Since $t_{0.025} = 2.110$ for d.f. = 17, the 95% confidence interval in this case is $-18 \pm 2.110(11.483)\sqrt{0.2159} = -18 \pm 11.3$ or $(-29.3, -6.7)$ pounds.

(c)   We assume normal populations with equal variances. The sample variances are not too different, and there are no obvious outliers in the data.

10.19   The summary statistics are:

Abused:	$n_1 = 52$,	$\overline{x} = 2.48$,	$s_1 = 1.94$
Non-abused:	$n_2 = 67$,	$\overline{y} = 1.57$,	$s_2 = 1.31$

Denote by μ_1 and μ_2 the population mean number of crimes for the "abused" and "non-abused" groups, respectively. Since the conjecture is that μ_1 is larger than μ_2, we formulate the hypotheses

$$H_0 : \mu_1 - \mu_2 = 0 \text{ versus } H_1 : \mu_1 - \mu_2 > 0\ .$$

Since the sample sizes are large, we employ the Z-test, and so, the test statistic is $Z = \dfrac{\overline{X} - \overline{Y}}{\sqrt{\dfrac{S_1^2}{n_1} + \dfrac{S_2^2}{n_2}}}$. Since H_1 is right-sided, the rejection region is of the form

$R : Z \geq c$. Using the summary statistics, we obtain

$$\overline{x} - \overline{y} = 2.48 - 1.57 = 0.91$$

$$\sqrt{\frac{s_1^2}{n_1} + \frac{s_2^2}{n_2}} = \sqrt{\frac{(1.94)^2}{52} + \frac{(1.31)^2}{67}} = 0.3130$$

Hence, the observed value of z is $Z = \dfrac{0.91}{0.3130} = 2.907$. The associated p-value is $P[Z \geq 2.907] = 0.00187$. This very low p-value signifies strong evidence in support of H_1.

10.21 (a) The summary statistics are:

Method 1: $\quad n_1 = 10$, $\qquad \bar{x} = 19.1$, $\qquad s_1 = 4.818$

Method 2: $\quad n_2 = 10$, $\qquad \bar{y} = 23.3$, $\qquad s_2 = 5.559$

Denote by μ_1 and μ_2 the population mean job times corresponding to Method 1 and Method 2, respectively. Since the conjecture is that μ_1 is smaller than μ_2, we formulate the hypotheses

$$H_0 : \mu_1 - \mu_2 = 0 \text{ versus } H_1 : \mu_1 - \mu_2 < 0 .$$

We assume normal populations with equal variance, so the test statistic is

$$T = \frac{\bar{X} - \bar{Y}}{s_{pooled}\sqrt{\dfrac{1}{n_1} + \dfrac{1}{n_2}}} \text{ with d.f.} = n_1 + n_2 - 2$$

Since H_1 is left-sided, the rejection region is $R : T \leq -t_{0.05} = -1.734$ for d.f. $= 10 + 10 - 2 = 18$. Using the summary statistics, we obtain

$$\bar{x} - \bar{y} = 19.1 - 23.3 = -4.2$$

$$s_{pooled} = \sqrt{\frac{(n_1 - 1)s_1^2 + (n_2 - 1)s_2^2}{n_1 + n_2 - 2}} = \sqrt{\frac{9(4.818)^2 + 9(5.559)^2}{18}} = 5.201$$

Hence, the observed value of t is given by

$$t = \frac{-4.2}{5.201\sqrt{\dfrac{1}{10} + \dfrac{1}{10}}} = -1.81,$$

which lies in R. So, H_0 is rejected at $\alpha = 0.05$. We conclude that the mean job time is significantly less for Method 1 than for Method 2.

(b) See part (a).

(c) A corresponding 95% confidence interval has the form

$$\left(\bar{X} - \bar{Y}\right) \pm t_{0.05/2}\, s_{pooled}\sqrt{\frac{1}{n_1} + \frac{1}{n_2}} .$$

Since $t_{0.025} = 2.101$ for d.f. $= 18$, using the calculations in part (a), we obtain that the 95% confidence interval in this case is

$$-4.2 \pm 2.101(2.326) = -4.2 \pm 4.89 \text{ or } (-9.09,\ 0.69).$$

10.23   The summary statistics are:

Isometric Method: $n_1 = 10$, $\bar{x} = 2.4$, $s_1 = 0.8$

Isotonic Method: $n_2 = 10$, $\bar{y} = 3.2$, $s_2 = 1.0$

(a)   Denote by μ_1 and μ_2 the population mean decrease in abdomen measurements under the Isometric Method and Isotonic Method, respectively. Since the intent is to demonstrate that the Isotonic Method is more effective (that is, μ_1 is smaller than μ_2), we formulate the hypotheses

$$H_0 : \mu_1 - \mu_2 = 0 \text{ versus } H_1 : \mu_1 - \mu_2 < 0 .$$

We assume normal populations with equal variance, so the test statistic is

$$T = \frac{\overline{X} - \overline{Y}}{S_{pooled}\sqrt{\dfrac{1}{n_1} + \dfrac{1}{n_2}}} \text{ with d.f.} = n_1 + n_2 - 2$$

Since H_1 is left-sided, the rejection region is $R : T \le -t_{0.05} = -1.734$ for d.f. $= 10 + 10 - 2 = 18$. Using the summary statistics, we obtain

$$\bar{x} - \bar{y} = 2.4 - 3.2 = -0.8$$

$$S_{pooled} = \sqrt{\frac{(n_1 - 1)s_1^2 + (n_2 - 1)s_2^2}{n_1 + n_2 - 2}} = \sqrt{\frac{9(0.8)^2 + 9(1.0)^2}{18}} = 0.906$$

Hence, the observed value of t is given by

$$t = \frac{-0.8}{0.906\sqrt{\dfrac{1}{10} + \dfrac{1}{10}}} = \frac{-0.8}{0.405} = -1.98,$$

which lies in R. So, H_0 is rejected at $\alpha = 0.05$. As such, the superiority of the Isotonic Method is demonstrated by the data. The rejection region and p-value are illustrated below (in that order):

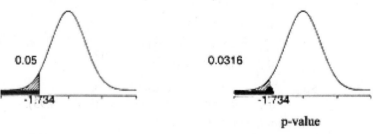

(b)   Using the calculations from part (a), a corresponding 95% confidence interval has the form

$$\left(\overline{X} - \overline{Y}\right) \pm t_{0.05/2} \, S_{pooled} \sqrt{\frac{1}{n_1} + \frac{1}{n_2}} = -0.8 \pm 2.101(0.405) = -0.8 \pm 0.85$$

or $(-1.65, 0.05)$ centimeters.

10.25   (a) The summary statistics are:

$$n_1 = 14, \qquad \bar{x} = 28.1, \qquad s_1 = 3.6$$
$$n_2 = 12, \qquad \bar{y} = 30.0, \qquad s_2 = 5.1$$

Assume normal populations with equal variances. A 98% confidence interval for $\mu_1 - \mu_2$ has the form

$$\left(\bar{X} - \bar{Y}\right) \pm t_{0.02/2}\, s_{pooled}\sqrt{\frac{1}{n_1} + \frac{1}{n_2}}$$

Note that $t_{0.01} = 2.492$ for d.f. $= 14 + 12 - 2 = 24$. Using the summary statistics, we obtain

$$\bar{x} - \bar{y} = -1.9$$

$$s_{pooled} = \sqrt{\frac{(n_1 - 1)s_1^2 + (n_2 - 1)s_2^2}{n_1 + n_2 - 2}} = \sqrt{\frac{13(3.6)^2 + 11(5.1)^2}{24}} = 4.352$$

So, the 98% confidence interval in this case is

$$-1.9 \pm 2.492(1.712) = -1.9 \pm 4.27 \quad \text{or} \quad (-6.17, 2.37).$$

  (b) The summary statistics are:

$$n_1 = 110, \qquad \bar{x} = 410, \qquad s_1 = 26$$
$$n_2 = 91, \qquad \bar{y} = 390, \qquad s_2 = 38$$

Since the sample sizes are large, neither the assumption of normal populations nor the assumption of equal variances is needed. A large sample 98% confidence interval for $\mu_1 - \mu_2$ is given by

$$\left(\bar{x} - \bar{y}\right) \pm z_{0.02/2}\sqrt{\frac{s_1^2}{n_1} + \frac{s_2^2}{n_2}}.$$

Observe that

$$z_{0.01} = 2.33$$
$$\bar{x} - \bar{y} = 410 - 390 = 20$$
$$\sqrt{\frac{s_1^2}{n_1} + \frac{s_2^2}{n_2}} = \sqrt{\frac{26^2}{110} + \frac{38^2}{90}} = 4.692$$

So, the 98% confidence interval becomes

$$20 \pm 2.33(4.692) = 20 \pm 10.93 \quad \text{or} \quad (9.07, 30.93).$$

  (c) The summary statistics are:

$$n_1 = 13, \qquad \bar{x} = 1.25, \qquad s_1 = 0.079$$
$$n_2 = 14, \qquad \bar{y} = 1.32, \qquad s_2 = 0.326$$

The sample sizes are both small. We assume that the populations are normal. The assumption of equal variances is not reasonable because $\frac{s_2}{s_1}$ is larger than 2. As such, we use the more conservative 98% confidence interval for $\mu_1 - \mu_2$ given by

$$(\bar{x}-\bar{y})\pm t^{*}_{0.02/2}\sqrt{\frac{s_1^2}{n_1}+\frac{s_2^2}{n_2}}\ ,\ \text{d.f.} = \text{smaller of } n_1-1 \text{ and } n_2-1.$$

Observe that

$$t^{*}_{0.01} = 2.764 \text{ for d.f.} = 12$$
$$\bar{x}-\bar{y} = -0.07$$
$$\sqrt{\frac{s_1^2}{n_1}+\frac{s_2^2}{n_2}} = \sqrt{\frac{(0.079)^2}{13}+\frac{(0.326)^2}{14}} = 0.0898$$

So, the 98% confidence interval becomes

$$-0.07\pm 2.681(0.0898) = -0.07\pm 0.241 \ \text{ or } \ (-0.311, 0.171).$$

(d) The summary statistics are:

$$n_1 = 65, \qquad \bar{x} = 75.6, \qquad s_1 = 18.1$$
$$n_2 = 55, \qquad \bar{y} = 62.5, \qquad s_2 = 6.8$$

The sample sizes are large, so we do not need the assumption of normal populations or equal variances. A 98% confidence interval for $\mu_1 - \mu_2$ is found (using the same method as in part (b)) to be

$$13.1\pm 2.33(2.425) = 13.1\pm 5.65 \ \text{ or } \ (7.45, 18.75).$$

10.27   The Minitab output is as follows:

```
Two sample T for WolfFwt vs WolMwt
```

	N	Mean	StDev	SE Mean
WolfFwt	8	73.9	10.1	3.6
WolMwt	11	91.9	12.4	3.7

```
97% CI for mu WolfFwt - mu WolMwt: ( -30.7,  -5.4)
T-Test mu WolfFwt = mu WolMwt (vs not =): T= -3.38 P=0.0036 DF=17
Both use Pooled StDev = 11.5
```

Since the p-value is between 0.025 and 0.01, we reject H_0.

10.29   We drew slips with α, β, and τ, so group 1 is {alpha, beta, tau}. (Answers may vary.)

10.31   We must be careful. It is likely that mothers who are warmer toward everyone have a much higher tendency to nurse their babies than mothers who have colder personalities. There are many other reasons as well.

10.33   (a) The $n=6$ paired differences $d = x - y$ are 2, 2, 4, 0, 1, -3.

Their mean and standard deviation are $\bar{d} = 1, \ s_D = 2.366$.

So, $t = \dfrac{1}{2.366/\sqrt{6}} = 1.035$.

(b) d.f. $= n-1 = 5$

10.35   It is a matched pair sample because there may be considerable variation of conditions in the different plants. The paired differences $d =$ (before – after) are $2, 1, -1, 2, 3, -1$. We assume these differences constitute a random sample from a normal distribution with mean δ. The null hypothesis of no change is $H_0 : \delta = 0$, and the alternative of more loss before than after is $H_1 : \delta > 0$. We calculate

$$\bar{d} = 1.0, \quad s_D = 1.673, \quad \tfrac{s_D}{\sqrt{6}} = 0.683, \quad t = \tfrac{1.0}{0.683} = 1.46, \text{ and d.f.} = 5.$$

With $\alpha = 0.05$, the rejection region is $R : T > t_{0.05} = 2.015$. Since the observed t is less than 2.015, H_0 is not rejected at $\alpha = 0.05$. As such, the claim of effectiveness of the safety program is not demonstrated.

10.37   This is a matched pair design because from each farm a pair of milk specimens are taken, and then one is treated with PC while the other is not.

(a)   The paired differences $d =$ (with PC – without PC) are $7, 6, -2, 10, 0, 8, 4$. From this, we obtain the following summary statistics:

$$n = 7, \quad \bar{d} = 4.714, \text{ and } s_D = 4.348$$

We assume that the population distribution of the D's is normal, and denote the population mean by δ. Because the conjecture is that the mean response with PC is higher than without PC (that is, $\delta > 0$), we formulate the hypotheses

$$H_0 : \delta = 0 \quad \text{versus} \quad H_1 : \delta > 0.$$

The test statistic is $T = \dfrac{\overline{D}}{S_D / \sqrt{n}}$, d.f. $= n - 1$. Since H_1 is right-sided and $\alpha = 0.05$, the rejection region is $R : T \geq t_{0.05} = 1.943$ (for d.f. $= 6$). Using the summary statistics, we see that the observed t is $t = \tfrac{4.714}{4.348/\sqrt{7}} = 2.87$, which lies in R. Hence, H_0 is rejected at $\alpha = 0.05$. Furthermore, the associated p-value is $P[T \geq 2.87] = 0.0142$. As such, we conclude that there is strong evidence in support of this conjecture. The rejection region and p-value are illustrated below (in that order):

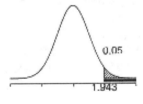 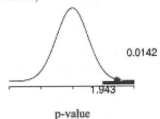

p-value

(b)   The corresponding 90% confidence interval for δ is given by

$$\bar{d} \pm t_{0.05}\, \tfrac{s_D}{\sqrt{n}} = 4.71 \pm 1.943 \left(\tfrac{4.35}{\sqrt{7}} \right) = 4.71 \pm 3.19 \text{ or } (1.52, 7.90).$$

10.39   This is a matched pair design. The paired differences $d = (A - B)$ are
$5, -1, 2, 3, 0, -1, 4, 3, -3$. From this, we obtain the following summary statistics:
$$n = 9, \ \overline{d} = 1.333, \text{ and } s_D = 2.693$$
We assume that the population distribution of these differences is normal, and
denote the population mean by δ. We formulate the hypotheses
$$H_0 : \delta = 0 \text{ versus } H_1 : \delta \neq 0.$$
The test statistic is $T = \dfrac{\overline{D}}{S_D / \sqrt{n}}$, d.f. $= n - 1$. Since H_1 is two-sided and
$\alpha = 0.05$, the rejection region is $R : |T| \geq t_{0.025} = 2.306$ (for d.f. = 8). Using the
summary statistics, we see that the observed t is $t = \frac{1.333}{2.693/\sqrt{9}} = 1.49$, which does
not lie in R. Hence, H_0 is not rejected at $\alpha = 0.05$, and so we conclude that the
difference is not significant at this level.

10.41   This is a matched pair design because the two plots in the same farm are alike
with respect to soil condition, rainfall, temperature, and other factors that affect
the yield. The paired differences $d = $ (yield of A $-$ yield of B) are
$5, 6, -2, 9, 0, 9, 7, 2$. From this, we obtain the following summary statistics:
$$n = 8, \ \overline{d} = 4.50, \text{ and } s_D = 4.11$$
We assume that the population distribution of these differences is normal, and
denote the population mean by δ.
(a)   In order to substantiate that A has a higher mean yield than B (that is,
$\delta > 0$), we formulate the hypotheses
$$H_0 : \delta = 0 \text{ versus } H_1 : \delta > 0.$$
The test statistic is $T = \dfrac{\overline{D}}{S_D / \sqrt{n}}$, d.f. $= n - 1$. Since H_1 is right-sided and
$\alpha = 0.05$, the rejection region is $R : T \geq t_{0.05} = 1.895$ (for d.f. = 7). Using
the summary statistics, we see that the observed t is $t = \frac{4.50}{4.11/\sqrt{8}} = 3.10$,
which lies in R. Hence, H_0 is rejected at $\alpha = 0.05$, and so we conclude
that A has a significantly higher mean yield than B.
(b)   At each farm, the assignment of the two plots to strain A and strain B
should be randomized. Label the two plots 1 and 2, and toss a coin – if a
Head turns up, assign plot 1 to strain A and plot 2 to strain B. If a Tail
turns up, the assign plot 1 to strain B and plot 2 to strain A. Repeat this
process for all the farms.

10.43   This is a matched pair design. The paired differences $d = $ (left $-$ right) are
$2, 3, 15, -2, -1, 1, -1, 7, 2, 10$. From this, we obtain the following summary
statistics: $n = 10, \ \overline{d} = 3.60, \text{ and } s_D = 5.461$
We assume that the population distribution of these differences is normal, and
denote the population mean by δ.
(a)   We formulate the hypotheses $H_0 : \delta = 0 \text{ versus } H_1 : \delta > 0$.

The test statistic is $T = \dfrac{\overline{D}}{S_D/\sqrt{n}}$, d.f. $= n-1$. Since H_1 is right-sided and $\alpha = 0.05$, the rejection region is $R: T \geq t_{0.05} = 1.833$ (for d.f. $= 9$). Using the summary statistics, we see that the observed t is $t = \frac{3.60}{5.461/\sqrt{10}} = 2.085$, which lies in R. Hence, H_0 is rejected at $\alpha = 0.05$.

(b) For $\alpha = 0.10$ and d.f. $= 9$, we have $t_{0.05} = 1.833$. So, the 90% confidence interval for δ is given by

$$\overline{d} \pm t_{0.05}\tfrac{s_D}{\sqrt{n}} = 3.6 \pm 1.833(1.727) = 3.6 \pm 3.17 \text{ or } (0.43, 6.77).$$

10.45 (a) Let p_A and p_B denote the population proportions of 'yes' for the ethnic groups A and B, respectively. We formulate the hypotheses:

$$H_0 : p_A = p_B \text{ versus } H_1 : p_A \neq p_B.$$

The test statistic is $Z = \dfrac{\hat{p}_1 - \hat{p}_2}{\sqrt{\hat{p}\hat{q}}\sqrt{\dfrac{1}{n_1}+\dfrac{1}{n_2}}}$, where $\hat{p}_1$ is identified as $\hat{p}_A$ and $\hat{p}_2$ is identified as $\hat{p}_B$. Since H_1 is two-sided and $\alpha = 0.05$, the rejection region is $R: |Z| \geq z_{0.05/2} = 1.96$ (illustrated below).

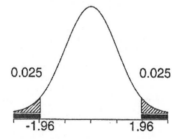

We calculate the following:

$$\hat{p}_1 = \tfrac{61}{100} = 0.61, \qquad \hat{p}_2 = \tfrac{31}{100} = 0.31$$

Pooled estimate $\hat{p} = \dfrac{n_1\hat{p}_1 + n_2\hat{p}_2}{n_1 + n_2} = \dfrac{100(0.61)+100(0.31)}{100+100} = 0.46$,

Observed z is $\dfrac{0.30}{\sqrt{(0.46)(0.54)}\sqrt{\dfrac{1}{100}+\dfrac{1}{100}}} = \dfrac{0.30}{0.0705} = 4.256$,

which lies in R. Hence, H_0 is rejected at $\alpha = 0.05$. Furthermore, the associated p-value $2P[Z \leq -3.911]$ is less than 0.0001.

(b) $\hat{p}_1 - \hat{p}_2 = 0.30$

Estimated S.E. $= \sqrt{\dfrac{\hat{p}_1\hat{q}_1}{n_1} + \dfrac{\hat{p}_2\hat{q}_2}{n_2}} = \sqrt{\dfrac{(0.61)(0.39)}{100} + \dfrac{(0.31)(0.69)}{100}} = 0.067$

So, a 95% confidence interval for $p_1 - p_2$ is given by

$$\left(\hat{p}_1 - \hat{p}_2\right) \pm z_{0.05/2} \sqrt{\frac{\hat{p}_1 \hat{q}_1}{n_1} + \frac{\hat{p}_2 \hat{q}_2}{n_2}} = 0.30 \pm 1.96(0.067) = 0.30 \pm 0.13$$

or $(0.169, 0.432)$.

10.47   Let p_1 and p_2 denote the probabilities of 'resistant' for the HRL and LRL groups, respectively. We formulate the hypotheses:

$$H_0 : p_1 = p_2 \quad \text{versus} \quad H_1 : p_1 < p_2.$$

The test statistic is $Z = \dfrac{\hat{p}_1 - \hat{p}_2}{\sqrt{\hat{p}\hat{q}} \sqrt{\dfrac{1}{n_1} + \dfrac{1}{n_2}}}$. Since H_1 is left-sided, the rejection

region is of the form $R : Z \le c$. We calculate the following:

$$\hat{p}_1 = \tfrac{15}{49} = 0.306, \qquad \hat{p}_2 = \tfrac{42}{54} = 0.778$$

Pooled estimate $\hat{p} = \dfrac{n_1 \hat{p}_1 + n_2 \hat{p}_2}{n_1 + n_2} = \dfrac{15 + 42}{49 + 54} = 0.553$,

Observed z is $\dfrac{0.306 - 0.778}{\sqrt{(0.553)(0.447)} \sqrt{\dfrac{1}{49} + \dfrac{1}{54}}} = -4.81$.

The associated p-value is $P[Z \le -4.81]$ is less than 0.0001. So, there is very strong evidence in support of H_1.

10.49   Let p_1 and p_2 denote the probabilities of being a chronic offender for abused and non-abused groups, respectively. We formulate the hypotheses:

$$H_0 : p_1 = p_2 \quad \text{versus} \quad H_1 : p_1 > p_2.$$

The test statistic is $Z = \dfrac{\hat{p}_1 - \hat{p}_2}{\sqrt{\hat{p}\hat{q}} \sqrt{\dfrac{1}{n_1} + \dfrac{1}{n_2}}}$. Since H_1 is right-sided and $\alpha = 0.01$,

the rejection region is $R : Z \ge z_{0.01} = 2.33$. We calculate the following:

$$\hat{p}_1 = \tfrac{21}{85} = 0.247, \qquad \hat{p}_2 = \tfrac{11}{120} = 0.092$$

Pooled estimate $\hat{p} = \dfrac{n_1 \hat{p}_1 + n_2 \hat{p}_2}{n_1 + n_2} = \dfrac{21 + 11}{85 + 120} = 0.156$,

Observed z is $\dfrac{0.247 - 0.092}{\sqrt{(0.156)(0.844)} \sqrt{\dfrac{1}{85} + \dfrac{1}{120}}} = 3.01$,

which lies in R. So, H_0 is rejected at $\alpha = 0.01$. Furthermore, the associated p-value is $P[Z \ge 3.01] = 0.0013$. So, there is very strong evidence in support of H_1. The rejection region and p-value are illustrated at the top of the next page (in that order):

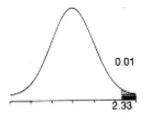
0.01

2.33

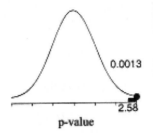
0.0013

2.58

p-value

10.51 Let p_1 and p_2 denote the probability of survival for the treated group (with carbolic acid) and the control group (without carbolic acid), respectively. We formulate the hypotheses:

$$H_0 : p_1 = p_2 \quad \text{versus} \quad H_1 : p_1 \neq p_2.$$

The test statistic is $Z = \dfrac{\hat{p}_1 - \hat{p}_2}{\sqrt{\hat{p}\hat{q}}\sqrt{\dfrac{1}{n_1} + \dfrac{1}{n_2}}}$. Since H_1 is two-sided and $\alpha = 0.05$,

the rejection region is $R : |Z| \geq z_{0.05/2} = 1.96$ (illustrated below).

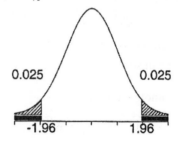

0.025   0.025

-1.96   1.96

We calculate the following:
$$\hat{p}_1 = \tfrac{34}{40} = 0.850, \qquad \hat{p}_2 = \tfrac{19}{35} = 0.543$$

Pooled estimate $\hat{p} = \dfrac{n_1\hat{p}_1 + n_2\hat{p}_2}{n_1 + n_2} = \dfrac{34 + 19}{40 + 35} = 0.707$,

Observed z is $\dfrac{0.850 - 0.543}{\sqrt{(0.707)(0.293)}\sqrt{\dfrac{1}{40} + \dfrac{1}{35}}} = 2.91$,

which lies in R. Hence, H_0 is rejected at $\alpha = 0.05$. Furthermore, the associated p-value $2P[Z \leq -2.91] = 2(0.0018) = 0.0036$. This means that H_0 would be rejected with α as small as 0.0036. As such, a difference in the survival rates is strongly demonstrated by the data.

10.53 Let p_1 and p_2 denote the population proportions of ≤ 8 hours of sleep for the age group $30 - 40$, and the age group $60 - 70$, respectively. We formulate the hypotheses: $H_0 : p_1 = p_2 \quad \text{versus} \quad H_1 : p_1 > p_2$.

The test statistic is $Z = \dfrac{\hat{p}_1 - \hat{p}_2}{\sqrt{\hat{p}\hat{q}}\sqrt{\dfrac{1}{n_1} + \dfrac{1}{n_2}}}$. Since H_1 is right-sided, the rejection

region is of the form $R : Z \geq c$. We calculate the following:

$$\hat{p}_1 = \tfrac{173}{250} = 0.692, \qquad \hat{p}_2 = \tfrac{120}{250} = 0.480$$

Pooled estimate $\hat{p} = \dfrac{n_1\hat{p}_1 + n_2\hat{p}_2}{n_1 + n_2} = \dfrac{293}{500} = 0.586$,

Observed z is $\dfrac{0.692 - 0.480}{\sqrt{(0.586)(0.414)}\sqrt{\dfrac{1}{250} + \dfrac{1}{250}}} = 4.81$

The associated p-value is $P[Z \geq 4.81]$ is less than 0.0002. So, there is very strong evidence in support of H_1.

10.55 (a) Let p_1 and p_2 denote the probability of having prominent wrinkles for smokers and non-smokers, respectively. We formulate the hypotheses:
$$H_0 : p_1 = p_2 \quad \text{versus} \quad H_1 : p_1 > p_2 .$$

The test statistic is $Z = \dfrac{\hat{p}_1 - \hat{p}_2}{\sqrt{\hat{p}\hat{q}}\sqrt{\dfrac{1}{n_1} + \dfrac{1}{n_2}}}$. Since H_1 is right-sided, the

rejection region is of the form $R : Z \geq c$. We calculate the following:

$$\hat{p}_1 = \tfrac{95}{150} = 0.633, \qquad \hat{p}_2 = \tfrac{103}{250} = 0.412$$

Pooled estimate $\hat{p} = \dfrac{n_1\hat{p}_1 + n_2\hat{p}_2}{n_1 + n_2} = \dfrac{95 + 103}{400} = 0.495$,

Observed z is $\dfrac{0.633 - 0.412}{\sqrt{(0.495)(0.505)}\sqrt{\dfrac{1}{150} + \dfrac{1}{250}}} = 4.28$

The associated p-value is $P[Z \geq 4.28]$ is less than 0.0002. So, there is very strong evidence in support of H_1.

(b) A direct causal relation between smoking and wrinkled skin cannot be readily concluded. Various psycho-physiological factors could influence both the smoking habit and the presence of wrinkled skin.

10.57 (a) Let p_1 and p_2 denote the probability of getting hepatitis for the 'vaccine' group and the 'placebo' group, respectively. We formulate the hypotheses:
$$H_0 : p_1 = p_2 \quad \text{versus} \quad H_1 : p_1 < p_2 .$$

The test statistic is $Z = \dfrac{\hat{p}_1 - \hat{p}_2}{\sqrt{\hat{p}\hat{q}}\sqrt{\dfrac{1}{n_1} + \dfrac{1}{n_2}}}$. Since H_1 is right-sided and

$\alpha = 0.01$, the rejection region is $R : Z \leq -z_{0.01} = -2.33$. We calculate the following:

$$\hat{p}_1 = \tfrac{11}{549} = 0.020, \qquad \hat{p}_2 = \tfrac{70}{534} = 0.131$$

Pooled estimate $\hat{p} = \dfrac{n_1\hat{p}_1 + n_2\hat{p}_2}{n_1 + n_2} = \dfrac{11+70}{1083} = 0.075$,

Observed z is $\dfrac{0.020 - 0.131}{\sqrt{(0.075)(0.925)}\sqrt{\dfrac{1}{549} + \dfrac{1}{534}}} = -6.93$,

which lies in R. Hence, H_0 is rejected at $\alpha = 0.01$. So, there is very strong evidence that the vaccine is effective.

(b) $\hat{p}_1 - \hat{p}_2 = 0.131 - 0.020 = 0.111$

Estimated S.E.$= \sqrt{\dfrac{\hat{p}_1\hat{q}_1}{n_1} + \dfrac{\hat{p}_2\hat{q}_2}{n_2}} = \sqrt{\dfrac{(0.020)(0.980)}{549} + \dfrac{(0.131)(0.869)}{534}}$
$$= 0.016$$

So, a 95% confidence interval for $p_1 - p_2$ is given by

$$\left(\hat{p}_1 - \hat{p}_2\right) \pm z_{0.05/2}\sqrt{\dfrac{\hat{p}_1\hat{q}_1}{n_1} + \dfrac{\hat{p}_2\hat{q}_2}{n_2}} = 0.111 \pm 1.96(0.016) = 0.111 \pm 0.031$$

or $(0.08, 0.14)$.

10.59 (a) $\hat{p}_E = \tfrac{78}{150} = 0.520, \qquad \hat{p}_H = \tfrac{39}{160} = 0.244$

$\hat{p}_E - \hat{p}_H = 0.276$

Estimated S.E.$= \sqrt{\dfrac{\hat{p}_1\hat{q}_1}{n_1} + \dfrac{\hat{p}_2\hat{q}_2}{n_2}} = \sqrt{\dfrac{(0.520)(0.480)}{150} + \dfrac{(0.244)(0.756)}{160}}$
$$= 0.053$$

So, a 95% confidence interval for $p_E - p_H$ is given by

$$\left(\hat{p}_E - \hat{p}_H\right) \pm z_{0.05/2}\sqrt{\dfrac{\hat{p}_E\hat{q}_E}{n_1} + \dfrac{\hat{p}_H\hat{q}_H}{n_2}} = 0.276 \pm 1.96(0.053) = 0.276 \pm 0.104$$

or $(0.17, 0.38)$.

(b) $\hat{p}_C = \tfrac{43}{200} = 0.215, \qquad \hat{p}_H = \tfrac{39}{160} = 0.244$

$\hat{p}_H - \hat{p}_C = 0.029$

Estimated S.E.$= \sqrt{\dfrac{\hat{p}_H\hat{q}_H}{n_1} + \dfrac{\hat{p}_C\hat{q}_C}{n_2}} = \sqrt{\dfrac{(0.244)(0.756)}{160} + \dfrac{(0.215)(0.785)}{200}}$
$$= 0.045$$

So, a 90% confidence interval for $p_H - p_C$ is given by

$$\left(\hat{p}_H - \hat{p}_C\right) \pm z_{0.10/2}\sqrt{\dfrac{\hat{p}_H\hat{q}_H}{n_1} + \dfrac{\hat{p}_C\hat{q}_C}{n_2}} = 0.029 \pm 1.645(0.045) = 0.029 \pm 0.074$$

or $(-0.045, 0.103)$.

(c) A 95% confidence interval for a population proportion is given by

$\hat{p} \pm 1.96 \sqrt{\dfrac{\hat{p}\hat{q}}{n}}$. Calculations for the individual groups are as follows:

Diabetes: $n = 160$, $\hat{p}_D = \frac{41}{160} = 0.256$

The confidence interval for p_D is given by:

$$0.256 \pm 1.96 \sqrt{\frac{(0.256)(0.744)}{160}} = 0.256 \pm 0.068 \quad \text{or} \quad (0.19, 0.32)$$

Heart condition: $n = 160$, $\hat{p}_H = \frac{39}{160} = 0.244$

The confidence interval for p_H is given by:

$$0.244 \pm 1.96 \sqrt{\frac{(0.244)(0.756)}{160}} = 0.244 \pm 0.067 \quad \text{or} \quad (0.18, 0.31)$$

Epilepsy: $n = 150$, $\hat{p}_E = \frac{78}{150} = 0.520$

The confidence interval for p_E is given by:

$$0.520 \pm 1.96 \sqrt{\frac{(0.520)(0.480)}{150}} = 0.520 \pm 0.080 \quad \text{or} \quad (0.44, 0.60)$$

Control: $n = 20$, $\hat{p}_C = \frac{43}{200} = 0.215$

The confidence interval for p_C is given by:

$$0.215 \pm 1.96 \sqrt{\frac{(0.215)(0.785)}{200}} = 0.215 \pm 0.057 \quad \text{or} \quad (0.16, 0.27)$$

10.61   For each case we are to test (at $\alpha = 0.05$)
$$H_0 : \mu_1 - \mu_2 = 0 \quad \text{versus} \quad H_1 : \mu_1 - \mu_2 \neq 0 \ .$$
The sample sizes are large ($n_1 = n_2 = 52$), so we employ the Z-test. Thus, the

test statistic is $Z = \dfrac{\overline{X} - \overline{Y}}{\sqrt{\dfrac{S_1^2}{n_1} + \dfrac{S_2^2}{n_2}}}$. Since H_1 is two-sided, the rejection region is

of the form $R : |Z| \geq z_{0.025} = 1.96$.

(a)   We have
$$|\overline{x} - \overline{y}| = 8$$
$$\sqrt{\frac{s_1^2}{n_1} + \frac{s_2^2}{n_2}} = \sqrt{\frac{20^2}{52} + \frac{28^2}{52}} = 4.772$$

Hence, the observed z is given by $|z| = \frac{8}{4.772} = 1.68$, which does not lie in R. Hence, H_0 is not rejected at $\alpha = 0.05$. So, the difference is not significant at $\alpha = 0.05$.

(b)   We have

$$|\bar{x} - \bar{y}| = 8$$

$$\sqrt{\frac{s_1^2}{n_1} + \frac{s_2^2}{n_2}} = \sqrt{\frac{12^2}{52} + \frac{15^2}{52}} = 2.664$$

Hence, the observed z is given by $|z| = \frac{8}{2.664} = 3.00$, which lies in R. Hence, H_0 is rejected at $\alpha = 0.05$. So, the difference is significant at $\alpha = 0.05$.

10.63 (a) Since the assertion is that $\mu_2 > \mu_1 + 10$, we formulate the hypotheses

$$H_0 : \mu_1 - \mu_2 = 10 \text{ versus } H_1 : \mu_2 - \mu_1 > 10 .$$

(b) Since the sample sizes $n_1 = 40$ and $n_2 = 48$ are large, we employ the Z-test, and so, the test statistic is $Z = \dfrac{(\bar{Y} - \bar{X}) - 10}{\sqrt{\dfrac{S_1^2}{n_1} + \dfrac{S_2^2}{n_2}}}$. Since H_1 is right-sided

and $\alpha = 0.10$, the rejection region is $R : Z \geq z_{0.10} = 1.28$.

(c) From the sample data, we calculate the value of the observed z to be

$$z = \frac{(27.84 - 16.21) - 10}{0.7721} = 2.11, \text{ which lies in } R. \text{ Hence, } H_0 \text{ is rejected at}$$

$\alpha = 0.10$. Furthermore, the associated p-value is $P[Z \geq 2.11] = 0.0174$, so the evidence is support of H_1 is strong.

10.65 (a) A large sample 90% confidence interval for $\mu_A - \mu_B$ is given by

$$(\bar{x} - \bar{y}) \pm z_{0.10/2} \sqrt{\frac{s_1^2}{n_1} + \frac{s_2^2}{n_2}} = (4.64 - 4.03) \pm 1.645 \sqrt{\frac{(1.25)^2}{55} + \frac{(1.82)^2}{58}}$$

$$= 0.61 \pm 0.48$$

or $(0.13, 1.09)$. We are 90% confident that μ_A is 0.13 to 1.09 hours longer than μ_B.

(b) The 95% confidence interval for μ_A is given by

$$\bar{x} \pm z_{0.05/2} \frac{s_1}{\sqrt{n_1}} = 4.64 \pm 1.96 \left(\frac{1.25}{\sqrt{55}} \right) = 4.64 \pm 0.33 \text{ or } (4.31, 4.97).$$

10.67 (a) We first obtain:

$$\bar{x} = 8 \qquad \qquad \bar{y} = 5$$

$$\sum (x_i - \bar{x})^2 = 14 \qquad \sum (y_i - \bar{y})^2 = 20$$

Consequently, the pooled variance is given by

$$s_{pooled}^2 = \frac{\sum (x_i - \bar{x})^2 + \sum (y_i - \bar{y})^2}{n_1 + n_2 - 2} = \frac{14 + 20}{8 + 5 - 2} = 4.857 .$$

(b)   The t-statistic is $T = \dfrac{(\overline{X}-\overline{Y})-2}{S_{pooled}\sqrt{\dfrac{1}{n_1}+\dfrac{1}{n_2}}}$ with d.f. $= n_1+n_2-2$. In the

present problem, we have $t = \dfrac{(8-5)-2}{\sqrt{4.857}\sqrt{\dfrac{1}{8}+\dfrac{1}{5}}} = 1.11$, with d.f. $= 7$.

10.69   Noting that for d.f. $= 11$, $t_{0.025} = 2.201$, and using the calculations in Exercise 10.68, we obtain the following 95% confidence interval for $\mu_1 - \mu_2$:

10.71   The summary statistics are:

City A: $n_1 = 75$, $\overline{x} = 37.8$, $s_1 = 6.8$
City B: $n_2 = 100$, $\overline{y} = 43.2$, $s_2 = 7.5$

(a)   Denote by μ_1 and μ_2 the population mean age for City A and City B, respectively. In order to demonstrate a difference between μ_1 and μ_2, we formulate the hypotheses
$$H_0 : \mu_1 - \mu_2 = 0 \ \text{ versus } \ H_1 : \mu_1 - \mu_2 \neq 0 \ .$$
Since the sample sizes are large, we employ the Z-test, and so, the test statistic is $Z = \dfrac{\overline{X}-\overline{Y}}{\sqrt{\dfrac{S_1^2}{n_1}+\dfrac{S_2^2}{n_2}}}$. Since H_1 is two-sided and $\alpha = 0.02$, the

rejection region is $R : |Z| \geq z_{0.02/2} = 2.33$. Using the summary statistics, we obtain
$$\overline{x}-\overline{y} = 37.8 - 43.2 = -5.4$$
$$\sqrt{\frac{S_1^2}{n_1}+\frac{S_2^2}{n_2}} = \sqrt{\frac{(6.8)^2}{75}+\frac{(7.5)^2}{100}} = 1.086$$

Hence, the observed value of z is $Z = \dfrac{-5.4}{1.086} = -4.97$, which lies in R.

Hence, H_0 is rejected at $\alpha = 0.02$, and so there is strong evidence that the mean ages are different.

(b) A large sample 98% confidence interval for $\mu_1 - \mu_2$ is given by

$$(\overline{x}-\overline{y}) \pm z_{0.02/2}\sqrt{\frac{S_1^2}{n_1}+\frac{S_2^2}{n_2}} = (37.8-43.2) \pm 2.33\sqrt{\frac{(6.8)^2}{75}+\frac{(7.5)^2}{100}} = -5.4 \pm 2.53$$

or $(-7.93, -2.87)$. We are 98% confident that the mean age for City B is 2.87 to 7.93 years higher than for City A.

(c) The individual 98% confidence intervals are given by:

$$\mu_1 : \bar{x} \pm 2.33 \frac{s_1}{\sqrt{n_1}} = 37.8 \pm 2.33 \left(\frac{6.8}{\sqrt{75}} \right) \text{ or } (35.97, 39.63) \text{ years}$$

$$\mu_2 : \bar{y} \pm 2.33 \frac{s_2}{\sqrt{n_2}} = 43.2 \pm 2.33 \left(\frac{7.5}{\sqrt{100}} \right) \text{ or } (41.45, 44.95) \text{ years}$$

10.73   (a)   Since the p-value is 0.067, we do not reject H_0 at $\alpha = 0.05$.

       (b)   A large sample 95% confidence interval for $\mu_1 - \mu_2$ is given by

$$(\bar{x} - \bar{y}) \pm z_{0.05/2} \sqrt{\frac{s_1^2}{n_1} + \frac{s_2^2}{n_2}} = (2.643 - 2.945) \pm 1.96 \sqrt{\frac{(0.554)^2}{15} + \frac{(0.390)^2}{20}}$$

$$= -0.302 \pm 1.96(0.1675) = -0.302 \pm 0.3283$$

or $(-0.6303, 0.0263)$.

       (c)   Denote by μ_1 and μ_2 the population mean CAS for female and male, respectively. In order to determine if μ_1 and μ_2 are significantly different, we test the hypotheses $H_0 : \mu_1 - \mu_2 = 0$ versus $H_1 : \mu_1 - \mu_2 \neq 0$ with $\alpha = 0.05$. We assume normal populations with equal variance, so the test statistic is

$$T = \frac{\bar{X} - \bar{Y}}{s_{pooled} \sqrt{\dfrac{1}{n_1} + \dfrac{1}{n_2}}} \text{ with d.f.} = n_1 + n_2 - 2$$

Since H_1 is two-sided and $\alpha = 0.05$, the rejection region is $R : |T| \geq t_{0.05/2} = 2.039$ (for d.f. $= 15 + 20 - 2 = 33$) – this is illustrated below. From the sample data, we calculate the following:
$\bar{x} - \bar{y} = 2.643 - 2.945 = -0.302$

$$s_{pooled} = \sqrt{\frac{(n_1 - 1)s_1^2 + (n_2 - 1)s_2^2}{n_1 + n_2 - 2}} = \sqrt{\frac{14(0.554)^2 + 19(0.390)^2}{33}} = 0.467.$$

As such, the value of the observed t is

$$t = \frac{-0.302}{0.3416(0.467)} = -1.89,$$

which does not lie in R. So, H_0 is not rejected at $\alpha = 0.05$.

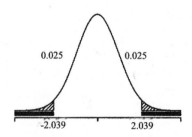

10.75   (a)   This is a matched pair design. From the data on the paired differences $d =$ (lab A − lab B), we obtain the following summary statistics:
$$n = 9, \quad \overline{d} = 9.667, \text{ and } s_D = 14.5$$
We assume that the population distribution of these differences is normal, and denote the population mean by δ. We formulate the hypotheses
$$H_0 : \delta = 0 \quad \text{versus} \quad H_1 : \delta \neq 0.$$
The test statistic is $T = \dfrac{\overline{D}}{S_D / \sqrt{n}}$, d.f. $= n-1$. Since H_1 is two-sided and $\alpha = 0.02$, the rejection region is $R : |T| \geq t_{0.01} = 2.896$ (for d.f. = 8). Using the summary statistics, we see that the observed t is $t = \frac{9.667}{14.5/\sqrt{9}} = 2.00$, which does not lie in R. Hence, H_0 is not rejected at $\alpha = 0.02$, and so we conclude that the difference is not significant at this level.

         (b)   For $\alpha = 0.10$ and d.f. = 8, we have $t_{0.05} = 1.860$. So, the 90% confidence interval for δ is given by
$$\overline{d} \pm t_{0.05} \tfrac{s_D}{\sqrt{n}} = 9.667 \pm 1.860\left(\tfrac{14.5}{\sqrt{9}}\right) = 9.667 \pm 8.999 \text{ or } (0.6789, 18.654).$$

10.77   For $\alpha = 0.05$ and d.f. = 9, we have $t_{0.025} = 2.262$. So, the 95% confidence interval for δ is given by
$$\overline{d} \pm t_{0.025} \tfrac{s_D}{\sqrt{n}} = 2.0 \pm 2.262\left(\tfrac{2.160}{\sqrt{10}}\right) = 2.0 \pm 1.55 \text{ or } (0.45, 3.55).$$

10.79   (a)   This is a case of independent random samples. We calculate the following summary statistics:

Path A:	$n_1 = 6$,	$\overline{x} = 12.5$,	$s_1 = 2.429$
Path B:	$n_2 = 6$,	$\overline{y} = 15.167$,	$s_2 = 2.317$

Denote by μ_1 and μ_2 the population mean travel times for Path A and Path B, respectively. Since the conjecture is that μ_1 and μ_2 are different, we formulate the hypotheses
$$H_0 : \mu_1 - \mu_2 = 0 \quad \text{versus} \quad H_1 : \mu_1 - \mu_2 \neq 0.$$
We assume normal populations with equal variance, so the test statistic is
$$T = \frac{\overline{X} - \overline{Y}}{S_{\text{pooled}} \sqrt{\dfrac{1}{n_1} + \dfrac{1}{n_2}}} \quad \text{with d.f.} = n_1 + n_2 - 2$$
Since H_1 is two-sided, the rejection region is $R : |T| \geq t_{0.05/2} = 2.228$ for d.f. $= 6 + 6 - 2 = 10$. Using the summary statistics, we obtain
$$\overline{x} - \overline{y} = -2.667$$
$$S_{\text{pooled}} = \sqrt{\frac{(n_1 - 1)s_1^2 + (n_2 - 1)s_2^2}{n_1 + n_2 - 2}} = \sqrt{\frac{5(2.429)^2 + 5(2.317)^2}{10}} = 2.374$$
Hence, the observed value of t is given by

$$t = \frac{-2.667}{2.374\sqrt{\dfrac{1}{6}+\dfrac{1}{6}}} = -1.947\,,$$

which does not lie in R. So, H_0 is not rejected at $\alpha = 0.05$. We conclude that the difference is not significant.

(b) Each driver could take both Path A and Path B. This would be a paired comparison that removes differences between drivers.

10.81 Let p_1 and p_2 denote the probabilities of rain from seeded clouds and non-seeded clouds, respectively. We formulate the hypotheses:

$$H_0: p_1 = p_2 \quad \text{versus} \quad H_1: p_1 < p_2\,.$$

The test statistic is $Z = \dfrac{\hat{p}_1 - \hat{p}_2}{\sqrt{\hat{p}\hat{q}}\sqrt{\dfrac{1}{n_1}+\dfrac{1}{n_2}}}$. Since H_1 is left-sided, the rejection region is of the form $R: Z \le c$. We calculate the following:

$$\hat{p}_1 = \tfrac{7}{50} = 0.140, \qquad \hat{p}_2 = \tfrac{43}{165} = 0.261$$

$$\text{Pooled estimate } \hat{p} = \frac{n_1\hat{p}_1 + n_2\hat{p}_2}{n_1 + n_2} = \frac{50}{215} = 0.233\,,$$

$$\text{Observed } z \text{ is } \frac{0.140 - 0.261}{\sqrt{(0.233)(0.767)}\sqrt{\dfrac{1}{50}+\dfrac{1}{165}}} = -1.77\,,$$

The associated p-value is $P[Z \le -1.77] = 0.0384$. So, H_0 would be rejected for α as small as 0.0384. As such, there is fairly strong evidence in support of the conjecture.

10.83 (a) Let p_1 and p_2 denote the population proportions of uremic and normal patients, respectively, who are allergic to the antibiotic. We formulate the hypotheses:

$$H_0: p_1 = p_2 \quad \text{versus} \quad H_1: p_1 > p_2\,.$$

The test statistic is $Z = \dfrac{\hat{p}_1 - \hat{p}_2}{\sqrt{\hat{p}\hat{q}}\sqrt{\dfrac{1}{n_1}+\dfrac{1}{n_2}}}$. Since H_1 is right-sided and $\alpha = 0.01$, the rejection region is $R: Z \ge z_{0.01} = 2.33$. We calculate the following:

$$\hat{p}_1 = \tfrac{38}{100} = 0.38, \qquad \hat{p}_2 = \tfrac{21}{100} = 0.21$$

$$\text{Pooled estimate } \hat{p} = \frac{n_1\hat{p}_1 + n_2\hat{p}_2}{n_1 + n_2} = \frac{38 + 21}{200} = 0.295\,,$$

$$\text{Observed } z \text{ is } \frac{0.38 - 0.21}{\sqrt{(0.295)(0.705)}\sqrt{\dfrac{1}{100}+\dfrac{1}{100}}} = 2.64\,,$$

which lies in R. So, H_0 is rejected at $\alpha = 0.01$. The associated p-value is $P[Z \geq 2.64] = 0.0041$. As such, there is strong evidence of a higher incidence of allergy in uremic patients. The rejection region and p-value are illustrated below (in that order):

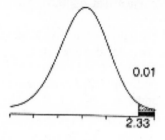

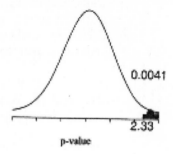

p-value

(b) $\hat{p}_1 - \hat{p}_2 = 0.38 - 0.21 = 0.17$

Estimated S.E. $= \sqrt{\dfrac{\hat{p}_1 \hat{q}_1}{n_1} + \dfrac{\hat{p}_2 \hat{q}_2}{n_2}} = \sqrt{\dfrac{(0.38)(0.62)}{100} + \dfrac{(0.21)(0.79)}{100}} = 0.063$

So, a 95% confidence interval for $p_1 - p_2$ is given by

$$(\hat{p}_1 - \hat{p}_2) \pm z_{0.05/2} \sqrt{\dfrac{\hat{p}_1 \hat{q}_1}{n_1} + \dfrac{\hat{p}_2 \hat{q}_2}{n_2}} = 0.17 \pm 1.96(0.063) = 0.17 \pm 0.12$$

or $(0.05, 0.29)$.

10.85   (a)   Since the p-value is 0.127, which is greater than 0.05, there is not strong enough evidence to reject H_0.

(b)   Enter the data into columns C3 and C4 of a Minitab worksheet. The output is as follows:

Two Sample T-Test and Confidence Interval
```
Two sample T for C3 vs C4

N       Mean      StDev     SE Mean
C3      8         73.9      10.1      3.6
C4      11        91.9      12.4      3.7

97% CI for mu C3 - mu C4: ( -30.3,   -5.8)
T-Test mu C3 = mu C4 (vs not =): T = -3.50
P = 0.0030 DF = 16
```
So, we would reject the null hypothesis $\alpha = 0.03$.

10.87   (a)   Denote by μ_1 and μ_2 the population mean for males at Lake Apopka and Lake Woodfruff, respectively. The summary statistics are:

Lake Apopka: $n_1 = 5$, $\bar{x} = 13.40$, $s_1 = 8.82$
Lake Woodruff: $n_2 = 9$, $\bar{y} = 50.44$, $s_2 = 27.13$

We should not pool the variances since the second standard deviation is more than 3 times the first.

(b) Using the conservative procedure, the 90% confidence interval for $\mu_1 - \mu_2$ has the form

$$\left(\overline{X} - \overline{Y}\right) \pm t_{\alpha/2} \sqrt{\frac{s_1^2}{n_1} + \frac{s_2^2}{n_2}} \; , \; \text{d.f.} = \text{smaller of } n_1 - 1, \, n_2 - 1.$$

Using the sample statistics, we obtain:
$$\overline{x} - \overline{y} = 13.40 - 50.44 = -37.04$$

$$\sqrt{\frac{s_1^2}{n_1} + \frac{s_2^2}{n_2}} = \sqrt{\frac{(8.82)^2}{5} + \frac{(27.13)^2}{9}} = 9.87$$

Since $t_{0.025} = 2.132$ for d.f. = 4, the 90% confidence interval in this case is
$-37.40 \pm (2.132)(9.87) = -37.40 \pm 21.04$ or $(-58.0, -16.0)$.

(c) The mean for males in the control group at Lake Woodruff is from 16 to 58 units larger than the mean for males at Lake Apopka.

10.89 Enter the data into columns C9 and C10 of a Minitab worksheet. The following is the output:

```
Two sample T for C9 vs C10

          N      Mean     StDev    SE Mean
C9       40     455.1      37.3        5.9
C10      40     429.1      41.0        6.5

95% CI for mu C9 - mu C10: ( 8.5,   43.4)
T-Test mu C9 = mu C10 (vs not =): T = 2.96   P = 0.0041   DF = 77
```

This was done without pooling the variance – note that since the ratio of the standard deviations is less than 1, one could have pooled the variance to get a slightly improved test statistic. Nonetheless, the p-value is 0.0041 with the more conservative estimate, so we reject H_0 and claim that the two means are different. Note also the 95% confidence interval is given.

10.91 Enter the data into two columns of a Minitab worksheet. The output is as follows:

```
Two sample T for Pre row vs Post row
             N      Mean     StDev    SE Mean
Pre row     81     725.9      89.3        9.9
Post row    81     673.3      83.0        9.2
95% CI for mu Pre row - mu Post row: ( 25.9,   79.4)
   T-Test mu Pre row = mu Post row (vs not =): T = 3.89
   P = 0.0001   DF = 159
```

There is a highly significant difference between mean pre row time and post row time.

Chapter 11

REGRESSION ANALYSIS I –
SIMPLE LINEAR REGRESSION

11.1   The points on the line $y = 2 + 3x$ for $x = 1$ and $x = 4$ are (1,5) and (4,14) respectively. The intercept is 2 and the slope is 3.

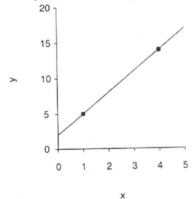

11.3   (a)   Predictor variable x = duration of training
            Response variable y = performance in skilled job
        (b)   Predictor variable x = average number of cigarettes smoked daily
            Response variable y = CO level in blood
        (c)   Predictor variable x = Humidity level in environment
            Response variable y = Growth rate of fungus

        (d)   Predictor variable x = Expenditures in promoting product
            Response variable y = Amount of product sales

11.5   The model is $Y = \beta_0 + \beta_1 x + e$, where $E(e) = 0$ and $sd(e) = \sigma$, so
        $\beta_0 = 7$, $\beta_1 = -5$, and $\sigma = 3$.

11.7 $Y = \beta_0 + \beta_1 x + e = 3 - 4x + e$, where $E(e) = 0$ and $sd(e) = \sigma$.
        (a)   At $x = 1$, $E(Y) = 3 - 4(1) = -1$ and $sd(Y) = sd(e) = 4$.
        (b)   At $x = 2$, $E(Y) = 3 - 4(2) = -5$ and $sd(Y) = sd(e) = 4$.

11.9   The straight line for the means of the model $Y = 7 + 2x + e$ is $y = 7 + 2x$. The
        graph of the line is shown below.

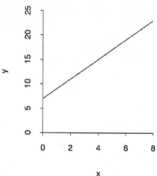

11.11   (a)   At $x = 4$, $E(Y) = \beta_0 + \beta_1(4) = 4 + 3(4) = 16$.
               At $x = 5$, $E(Y) = \beta_0 + \beta_1(5) = 4 + 3(5) = 19$.
        (b)   No, only the mean is larger. By chance the error e at $x = 5$, which has
               standard deviation 4, could be quite negative and/or the error at $x = 4$ very
               large.

11.13   (a)   The scatter diagram is shown below.

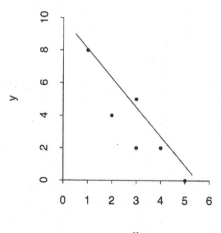

        (b)   The computations needed to calculate $\bar{x}$, $\bar{y}$, S_{xx}, S_{xy}, and S_{yy} are provided in the following table:

x	y	$x-\bar{x}$	$y-\bar{y}$	$(x-\bar{x})(y-\bar{y})$	$(x-\bar{x})^2$	$(y-\bar{y})^2$
1	8	−2	4.5	−9.0	4	20.25
2	4	−1	0.5	−0.5	1	0.25
3	5	0	1.5	0	0	2.25
3	2	0	−1.5	0	0	2.25
4	2	1	−1.5	−1.5	1	2.25
5	0	2	−3.5	−7.0	4	12.25
Total	18	21	0	0	−18.0	10	39.50

So, we have

$$\bar{x} = \frac{\sum x}{n} = \frac{18}{6} = 3 \qquad S_{xx} = \sum(x-\bar{x})^2 = 10$$

$$\bar{y} = \frac{\sum y}{n} = \frac{21}{6} = 3.5 \qquad S_{xy} = \sum(x-\bar{x})(y-\bar{y}) = -18.0$$

$$S_{yy} = \sum(y-\bar{y})^2 = 39.50$$

(c) $\hat{\beta}_1 = \dfrac{S_{xy}}{S_{xx}} = -\dfrac{18}{10} = -1.8, \quad \hat{\beta}_0 = \bar{y} - \hat{\beta}_1\bar{x} = 3.5 - (-1.8)(3) = 8.9$

(d) The fitted line is $\hat{y} = 8.9 - 1.8x$, which is graphed in part (a).

11.15 (a) The residuals and their sum are calculated in the following table:

x	y	$\hat{y} = 8.9 - 1.8x$	$\hat{e} = y - \hat{y}$	$(y-\hat{y})^2$
1	8	7.1	0.9	0.81
2	4	5.3	−1.3	1.69
3	5	3.5	1.5	2.25
3	2	3.5	−1.5	2.25
4	2	1.7	0.3	0.09
5	0	−0.1	0.1	0.01
Total			0.0	7.10

So, $\sum(y-\hat{y}) = 0.0$, SSE = 7.10.

(b) SSE = Sum of squares residuals = 7.10

$$\text{SSE} = S_{yy} - \frac{S_{xy}^2}{S_{xx}} = 39.5 - \frac{(-18.0)^2}{10} = 7.10 \quad \text{(Check this using the}$$

calculations in Exercise 11.13.)

(c) $S^2 = \dfrac{\text{SSE}}{n-2} = \dfrac{7.10}{6-2} = 1.775$

11.17   (a)   The computations needed to calculate $\bar{x}$, $\bar{y}$, S_{xx}, S_{xy}, and S_{yy} are provided in the following table:

x	y	$x-\bar{x}$	$y-\bar{y}$	$(x-\bar{x})(y-\bar{y})$	$(x-\bar{x})^2$	$(y-\bar{y})^2$
0	4	−4	−2	8	16	4
2	3	−2	−3	6	4	9
4	6	0	0	0	0	0
6	8	2	2	4	4	4
8	9	4	3	12	16	9
Total	20	30	0	0	30	40	26

So, we have

$$\bar{x} = \frac{\sum x}{n} = \frac{20}{5} = 4 \qquad S_{xx} = \sum(x-\bar{x})^2 = 40$$

$$\bar{y} = \frac{\sum y}{n} = \frac{30}{5} = 6 \qquad S_{xy} = \sum(x-\bar{x})(y-\bar{y}) = 30$$

$$S_{yy} = \sum(y-\bar{y})^2 = 26$$

(b) $\hat{\beta}_1 = \dfrac{S_{xy}}{S_{xx}} = \dfrac{30}{40} = 0.75, \quad \hat{\beta}_0 = \bar{y} - \hat{\beta}_1\bar{x} = 6 - (0.75)(4) = 3.00$

(c)   The fitted line is $\hat{y} = 3.00 - 0.75x$.

11.19   (a) $\hat{\beta}_1 = \dfrac{S_{xy}}{S_{xx}} = \dfrac{3.25}{28.20} = 0.1152, \quad \hat{\beta}_0 = \bar{y} - \hat{\beta}_1\bar{x} = 5.2 - (0.1152)(1.4) = 5.04$

The fitted line is $\hat{y} = 5.04 + 0.115x$.

(b) $\text{SSE} = S_{yy} - \dfrac{S_{xy}^2}{S_{xx}} = 2.01 - \dfrac{(3.25)^2}{28.2} = 1.635$

(c) $S^2 = \dfrac{\text{SSE}}{n-2} = \dfrac{1.635}{18} = 0.091$

11.21   We first calculate the means and sums of squares and products:

$$\bar{y} = \frac{\sum y}{n} = \frac{889}{7} = 127, \qquad \bar{x} = \frac{\sum x}{n} = \frac{520}{7} = 74.286$$

$$\sum_{i=1}^{7} y_i^2 = 113,237, \qquad \sum_{i=1}^{7} x_i^2 = 39,328, \qquad \sum_{i=1}^{7} x_i y_i = 66,392$$

Thus,

$$S_{xx} = \sum_{i=1}^{7} x_i^2 - \left(\sum_{i=1}^{7} x_i \right)^2 / 7 = 39,328 - (520)^2 / 7 = 699.43$$

$$S_{xy} = \sum_{i=1}^{7} x_i y_i - \left(\sum_{i=1}^{7} x_i \right)\left(\sum_{i=1}^{7} y_i \right) / 7 = 66,392 - (520)(889)/7 = 352$$

$$S_{yy} = \sum_{i=1}^{7} y_i^2 - \left(\sum_{i=1}^{7} y_i \right)^2 / 7 = 113,237 - (889)^2 / 7 = 334$$

(a) $\hat{\beta}_1 = \dfrac{S_{xy}}{S_{xx}} = \dfrac{352}{669.43} = 0.5033,$

$\hat{\beta}_0 = \bar{y} - \hat{\beta}_1 \bar{x} = 127 - (0.5033)(74.286) = 89.61$

The fitted line is $\hat{y} = 89.61 + 0.5033x$.

(b) $\text{SSE} = S_{yy} - \dfrac{S_{xy}^2}{S_{xx}} = 334 - \dfrac{(352)^2}{699.43} = 156.85$

(c) $S^2 = \dfrac{\text{SSE}}{n-2} = \dfrac{156.85}{5} = 31.37$

(d) No, the two should be inverses of each other (the graphs of which are reflections over the $y = x$ line). Hence, they will not have the same equation <u>unless</u> they are both identical to $y = x$.

11.23 At $\bar{x}$, $\hat{y} = \hat{\beta}_0 + \hat{\beta}_1 \bar{x} = \underbrace{\bar{y} - \hat{\beta}_1 \bar{x}}_{=\hat{\beta}_0} + \hat{\beta}_1 \bar{x} = \bar{y}$

11.25 (a) The computations needed to calculate $\bar{x}$, $\bar{y}$, S_{xx}, S_{xy}, and S_{yy} are provided in the following table:

x	y	$x - \bar{x}$	$y - \bar{y}$	$(x - \bar{x})(y - \bar{y})$	$(x - \bar{x})^2$	$(y - \bar{y})^2$
0	5	−3	2	−6	9	4
1	4	−2	1	−2	4	1
6	1	3	−2	−6	9	4
3	3	0	0	0	0	0
5	2	2	−1	−2	4	1
Total	15	15	0	0	−16	26	10

So, we have

$$\bar{x} = \frac{\sum x}{n} = \frac{15}{5} = 3 \qquad S_{xx} = \sum (x - \bar{x})^2 = 26$$

$$\bar{y} = \frac{\sum y}{n} = \frac{15}{5} = 3 \qquad S_{xy} = \sum (x - \bar{x})(y - \bar{y}) = -16$$

$$S_{yy} = \sum (y - \bar{y})^2 = 10$$

$$\hat{\beta}_1 = \frac{S_{xy}}{S_{xx}} = -\frac{16}{26} = -0.615, \qquad \hat{\beta}_0 = \bar{y} - \hat{\beta}_1 \bar{x} = 3 - (-0.615)(3) = 4.845$$

$$\text{SSE} = S_{yy} - \frac{S_{xy}^2}{S_{xx}} = 10 - \frac{(-16)^2}{26} = 0.1538$$

$$S^2 = \frac{\text{SSE}}{n-2} = \frac{0.1538}{3} = 0.0513$$

(b)   We test the hypotheses: $H_0 : \beta_1 = 0$ versus $H_1 : \beta_1 \neq 0$

Since H_1 is two –sided and $\alpha = 0.05$, the rejection region is

$R : |T| > t_{0.025} = 3.182$ (for d.f. = 3). The value of the observed t is

$$t = \frac{\hat{\beta}_1 - 0}{s / \sqrt{S_{xx}}} = \frac{-0.615}{\sqrt{0.0513/26}} = -13.8,$$

which lies in R. Hence, H_0 is rejected at $\alpha = 0.05$.

(c)   The expected value is estimated by $4.845 - 0.615(2.5) = 3.308$. Since

$$s \sqrt{\frac{1}{n} + \frac{(2.5 - \bar{x})^2}{S_{xx}}} = \sqrt{0.0513} \sqrt{\frac{1}{5} + \frac{(2.5-3)^2}{26}} = 0.104$$

and the upper 0.05 point of the t with d.f. = 3 is 2.353, the 90% confidence interval for the expected y value is given by

$$3.308 \pm 2.353(0.104) \quad \text{or} \quad (3.06, 3.55).$$

11.27   Since $t_{0.025} = 3.182$ for d.f. = 3, a 95% confidence interval for β_1 is given by

$$\hat{\beta}_1 \pm 3.182 \frac{s}{\sqrt{S_{xx}}} = -0.615 \pm 3.182 \sqrt{\frac{0.0513}{26}}$$

or $(-0.756, -0.474)$.

11.29   (a) & (b) Using Minitab, we have:

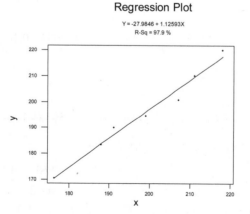

Regression Plot

Y = -27.9846 + 1.12593X
R-Sq = 97.9 %

Note the equation of the least squares regression line is $y = 1.12593x - 27.9846$. This could have been computed by hand using the same calculations used in exercises from the previous section.

(c) First, note that $t_{0.025} = 2.571$ for d.f. = 5. In order to determine the confidence interval for $\hat{\beta}_1$, we need the following computations:

x	y	$x-\bar{x}$	$y-\bar{y}$	$(x-\bar{x})(y-\bar{y})$	$(x-\bar{x})^2$	$(y-\bar{y})^2$
183.5	188	−12.2	−10.6	129.32	148.84	112.36
190	191.2	−5.7	−7.4	42.18	32.49	54.76
170.5	176.2	−25.2	−22.4	564.48	635.04	501.76
200.8	207	5.1	8.4	42.84	26.01	70.56
210.2	211	14.5	12.4	179.80	210.25	153.76
194.6	199	−1.1	0.4	−0.44	1.21	0.16
220	218	24.3	19.4	471.40	590.49	376.36
Total				1429.6	1644.3	1269.7

So, we have

$$\bar{x} = \frac{\sum x}{7} = 195.7 \qquad S_{xx} = \sum (x-\bar{x})^2 = 1644.3$$

$$\bar{y} = \frac{\sum y}{7} = 198.6 \qquad S_{xy} = \sum (x-\bar{x})(y-\bar{y}) = 1429.6$$

$$S_{yy} = \sum (y-\bar{y})^2 = 1269.7$$

Thus, we have

$$\text{SSE} = S_{yy} - \frac{S_{xy}^2}{S_{xx}} = 26.766 \qquad S^2 = \frac{\text{SSE}}{n-2} = \frac{26.766}{5} = 5.3532.$$

Therefore, $\dfrac{s}{\sqrt{S_{xx}}} = \dfrac{2.314}{\sqrt{1644.3}} = 0.0571$. Also, $\hat{\beta}_1 = \dfrac{S_{xy}}{S_{xx}} = 0.8694$.

So, a 95% confidence interval for β_1 is given by

$$\hat{\beta}_1 \pm 2.571 \frac{s}{\sqrt{S_{xx}}} = 0.8694 \pm 2.571(0.0571) = 0.8694 \pm 0.1468$$

or $(0.7226, 1.016)$.

11.31 (a) $\hat{\beta}_1 = \dfrac{S_{xy}}{S_{xx}} = \dfrac{6.7}{4.2} = 1.595$, $\hat{\beta}_0 = \bar{y} - \hat{\beta}_1 \bar{x} = 4.6 - (1.595)(1.1) = 2.846$

The fitted line is then given by $\hat{y} = 2.85 + 1.60x$.

(b) $\text{SSE} = S_{yy} - \dfrac{S_{xy}^2}{S_{xx}} = 12.2 - \dfrac{(6.7)^2}{4.2} = 1.512$ $\qquad S^2 = \dfrac{\text{SSE}}{n-2} = \dfrac{1.512}{13} = 0.116$

(c) We test the hypotheses: $H_0 : \beta_1 = 1.3$ versus $H_1 : \beta_1 > 1.3$

Since H_1 is right-sided and $\alpha = 0.05$, the rejection region is

$R : T > t_{0.05} = 1.771$ (for d.f. = 13). The value of the observed t is

$$t = \frac{\hat{\beta}_1 - 1.3}{s / \sqrt{S_{xx}}} = \frac{1.595 - 1.3}{\sqrt{0.116 / 4.2}} = 1.775,$$

which lies in R. Hence, H_0 is rejected at $\alpha = 0.05$.

11.33   (a)   The model is $Y = \beta_0 + \beta_1 x + e$ and the fit suggested by the data is

$\hat{y} = 994 + 0.10373x$ with $sd(e) = \hat{\sigma} = 299.4$.

Note that the r^2 is only 0.302. This means that only 30.2% of the variability in the data is explained by the model (refer to Exercise 11.44).

(b)   The t-ratio on the computer output is the t-statistic for testing that the coefficient is zero. Since the t-ratio for the x term is 3.48 with p-value 0.002, we reject H_0: $\beta_0 = 0$ at $\alpha = 0.05$.

11.35   (a)   The model is $Y = \beta_0 + \beta_1 x + e$ and the fit suggested by the data is

$\hat{y} = 0.3381 + 0.83099x$ with $sd(e) = \hat{\sigma} = 0.1208$.

(b)   Since the t-ratio for the x term is 9.55 with p-value less than 0.0001, we reject H_0: $\beta_1 = 0$ at $\alpha = 0.05$. As such, the x term is needed in the model.

11.37   (a)   Using Minitab, we find that the fitted line plot is as follows:

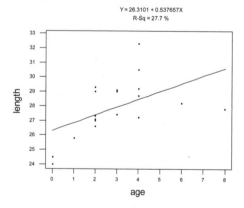

(b)   Enter the data into a Minitab worksheet. The output is as follows:

Regression Analysis

```
The regression equation is
length = 26.3 + 0.538 age


Predictor          Coef         StDev          T          P
Constant        26.3101        0.7356      35.77      0.000
age              0.5377        0.2105       2.55      0.021

S = 1.722        R-Sq = 27.7%        R-Sq(adj) = 23.5%
```

```
Analysis of Variance

Source          DF      SS          MS          F         P
Regression      1       19.353      19.353      6.52      0.021
Residual Error  17      50.437      2.967
Total           18      69.789
```

Look in the age row – the *p*-value of 0.021 is the result of the hypothesis test of the slope at the 95% level. Here, we reject H_0 in favor of claiming there is linear relationship between the two variables.

(c) & (d) One can proceed by hand as in other exercises/examples. We, however choose to use Minitab to obtain the following two 90% confidence intervals corresponding to age $x = 4$:

```
Predicted Values
Fit   StDev Fit           90.0% CI                90.0% PI
28.461        0.453   (   27.673,   29.249)   (   25.362,   31.559)
```

11.39 $r^2 = \dfrac{S_{xy}^2}{S_{xx}S_{yy}} = \dfrac{(9.4)^2}{(43.2)(6.7)} = 0.305$

11.41 $r^2 = \dfrac{S_{xy}^2}{S_{xx}S_{yy}} = \dfrac{(16)^2}{(9.2)(49)} = 0.568$

11.43 By Exercise 11.28, we have $S_{xx} = 10$, $S_{yy} = 7.0$, $S_{xy} = 5.06$. So,

(a) Proportion of variance explained $r^2 = \dfrac{S_{xy}^2}{S_{xx}S_{yy}} = 0.3658$

(b) $r = \dfrac{S_{xy}}{\sqrt{S_{xx}S_{yy}}} = 0.6047$

11.45 Proportion explained $= r^2 = 0.799$

11.47 (a) $\hat{\beta}_1 = \dfrac{S_{xy}}{S_{xx}}$, so multiplying by $\dfrac{S_{xx}}{S_{xx}}$

$r = \dfrac{S_{xy}}{\sqrt{S_{xx}S_{yy}}} = \dfrac{S_{xx}}{S_{xx}} \dfrac{S_{xy}}{\sqrt{S_{xx}}} \dfrac{1}{\sqrt{S_{yy}}} = \hat{\beta}_1 \dfrac{S_{xx}}{\sqrt{S_{xx}}} \dfrac{1}{\sqrt{S_{yy}}} = \hat{\beta}_1 \dfrac{\sqrt{S_{xx}}}{\sqrt{S_{yy}}}$

(b) $\text{SSE} = S_{yy} - \dfrac{S_{xy}^2}{S_{xx}} = S_{yy} - \dfrac{S_{xy}^2}{S_{xx}} \dfrac{S_{yy}}{S_{yy}} = S_{yy}\left(1 - \dfrac{S_{xy}^2}{S_{xx}S_{yy}}\right) = S_{yy}(1 - r^2)$

11.49 The product $x = $ (leaf length) $\times$ (leaf width) is the area of a rectangle that contains the leaf. It should be larger than the leaf, so the slope should be less than one.

11.51   (a)   The residuals and their sum are calculated in the following table:

x	y	$\hat{y}$	$\hat{e}=y-\hat{y}$	$(y-\hat{y})^2$
1	9	7.39	1.61	2.592
1	7	7.39	−0.39	0.152
1	8	7.39	0.61	0.372
2	10	11.05	−1.05	1.103
3	15	14.70	0.30	0.090
3	12	14.70	−2.70	7.290
4	19	18.36	0.646	0.410
5	24	22.01	1.99	3.960
5	21	22.01	−1.01	1.020
Total				16.989

So, SSE $=16.989$.

(b)   SSE = Sum of squares residuals = 16.989

$$\text{SSE} = S_{yy} - \frac{S_{xy}^2}{S_{xx}} = 304.889 - \frac{(78.778)^2}{21.556} = 16.989$$

(c) $S^2 = \dfrac{\text{SSE}}{n-2} = \dfrac{16.989}{9-2} = 2.427$

11.53   (a) $\hat{\beta}_1 = \dfrac{S_{xy}}{S_{xx}} = -\dfrac{12.4}{5.6} = -2.214, \quad \hat{\beta}_0 = \bar{y} - \hat{\beta}_1\bar{x} = 54.8 + (2.214)(8.3) = 73.18$

The fitted line is then given by $\hat{y} = 73.18 - 2.214x$.

Also, $\text{SSE} = S_{yy} - \dfrac{S_{xy}^2}{S_{xx}} = 38.7 - \dfrac{(12.4)^2}{5.6} = 11.24$

$$S^2 = \frac{\text{SSE}}{n-2} = \frac{11.24}{13} = 0.8648$$

(b)   We test the hypotheses: $H_0 : \beta_1 = -2$ versus $H_1 : \beta_1 < -2$

Since H_1 is left-sided and $\alpha = 0.05$, the rejection region is

$R : T < -t_{0.05} = -1.771$ (for d.f. = 13). The value of the observed t is

$$t = \frac{\hat{\beta}_1 - (-2)}{s/\sqrt{S_{xx}}} = \frac{-2.214 + 2}{\sqrt{0.8648/5.6}} = -1.289,$$

which does not lie in R. Hence, H_0 is not rejected at $\alpha = 0.05$.

(c)   A 95% confidence interval (for $x^* = 10$) of the expected response 51.04 is given by

$$51.04 \pm 2.160(0.93)\sqrt{\frac{1}{15} + \frac{(10-8.3)^2}{5.6}} \quad \text{or} \quad (49.51, 52.57).$$

11.55   (a)   The scatter diagram is given by

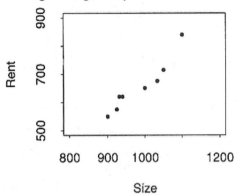

We first calculate the means and sums of squares and products:

$$\bar{x} = \frac{\sum x}{8} = \frac{7880}{8} = 985, \qquad \bar{y} = \frac{\sum y}{8} = \frac{5245}{8} = 655.63$$

$$\sum_{i=1}^{8} x_i^2 = 7,797,438, \qquad \sum_{i=1}^{8} y_i^2 = 3,496,875, \qquad \sum_{i=1}^{8} x_i y_i = 5,209,540$$

Thus,

$$S_{xx} = \sum_{i=1}^{8} x_i^2 - \left(\sum_{i=1}^{8} x_i \right)^2 / 8 = 7,797,438 - (7880)^2 / 8 = 35,638$$

$$S_{xy} = \sum_{i=1}^{8} x_i y_i - \left(\sum_{i=1}^{8} x_i \right)\left(\sum_{i=1}^{8} y_i \right) / 8 = 5,209,540 - (7880)(5245) / 8 = 43,215$$

$$S_{yy} = \sum_{i=1}^{8} y_i^2 - \left(\sum_{i=1}^{8} y_i \right)^2 / 8 = 3,496,875 - (5245)^2 / 8 = 58,121.88$$

Consequently, we have

$$\hat{\beta}_1 = \frac{S_{xy}}{S_{xx}} = \frac{43,215}{35,638} = 1.2126,$$

$$\hat{\beta}_0 = \bar{y} - \hat{\beta}_1 \bar{x} = 655.63 - (\tfrac{43,215}{35,638})(985) = -538.79$$

The fitted line is $\hat{y} = -538.79 + 1.2126x$.
Furthermore,

$$\text{SSE} = S_{yy} - \frac{S_{xy}^2}{S_{xx}} = 58,121.88 - \frac{(43,215)^2}{35,638} = 5,718.93$$

$$S^2 = \frac{\text{SSE}}{n-2} = \frac{5,718.93}{6} = 953.155, \text{ so that } S = 30.87.$$

(b)   We test the hypotheses: $H_0 : \beta_1 = 0$ versus $H_1 : \beta_1 > 0$
Since H_1 is right-sided and $\alpha = 0.05$, the rejection region is
$R : T > t_{0.05} = 1.943$ (for d.f. = 6). The value of the observed t is

$$t = \frac{\hat{\beta}_1 - 0}{s/\sqrt{S_{xx}}} = \frac{1.2126}{30.87\sqrt{35,638}} = 7.42,$$

which lies in R. Hence, H_0 is rejected at $\alpha = 0.05$. This implies that the mean rent y increases with size x.

(c) The expected increase is β_1. A 95% confidence interval is calculated as

$$1.2126 \pm 2.447\left(\frac{30.87}{\sqrt{35,638}}\right) = 1.2126 \pm 0.400 \quad \text{or} \quad (0.813, \ 1.613).$$

(d) For a specific apartment of size $x = 1025$, a 95% prediction interval is given by

$$\left[-538.79 + 1.2126(1025)\right] \pm 2.447(30.87)\sqrt{1 + \frac{1}{8} + \frac{(1025 - 985)^2}{35,638}}$$

$$= 704.125 \pm 81.704$$

or $(622.4, \ 785.8)$.

11.57 (a) & (b) The scatter diagram and fitted line are shown below:

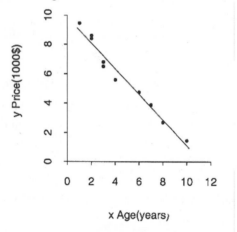

$$\bar{x} = \frac{\sum x}{10} = 4.4 \qquad S_{xx} = \sum (x - \bar{x})^2 = 50.24$$

$$\bar{y} = \frac{\sum y}{10} = 10.189 \qquad S_{xy} = \sum (x - \bar{x})(y - \bar{y}) = -78.256$$

$$S_{yy} = \sum (y - \bar{y})^2 = 136.409$$

So, the fitted line is $\hat{y} = 17.11 - 1.558x$.

(c) Since $t_{0.025} = 2.365$ for d.f. = 7, a 95% confidence interval for β_1 is given

by $\hat{\beta}_1 \pm 2.306 \dfrac{s}{\sqrt{S_{xx}}} = -1.558 \pm 2.365\left(\dfrac{1.438}{\sqrt{50.24}}\right) = -1.558 \pm 0.480$

or $(-2.038, \ -1.078)$.

11.59 $r = \dfrac{S_{xy}}{\sqrt{S_{xx}S_{yy}}} = \dfrac{-78.2556}{\sqrt{(50.24)(136.409)}} = -0.8939$. Since $r^2 = 0.945$ is the

proportion of variance explained by the linear regression of y on x, the fit appears to be adequate.

11.61   (a) $\hat{\beta}_0 = -1.071,\quad \hat{\beta}_1 = 2.7408$

(b)   SSE $= 63.65$

(c)   Estimated S.E. of $\hat{\beta}_0$ is 2.751,

Estimated S.E. of $\hat{\beta}_1$ is 0.4411.

(d)   For testing $H_0 : \beta_0 = 0,\ \ t = -0.39$.

For testing $H_0 : \beta_1 = 0,\ \ t = 6.21$.

(e) $r^2 = 0.828$

(f)   The decomposition of the total sum of squares is

$$\underbrace{\text{Total}}_{S_{yy}}\ =\ \underbrace{\text{Explained by linear relation}}_{\dfrac{S_{xy}^2}{S_{xx}}}\ +\ \underbrace{\text{Error}}_{\text{SSE}}$$

$$307.90\ =\ \qquad\qquad 307.25 \qquad\qquad +\ \ 63.65$$

11.63   The Minitab output is on the next page.

(a)   From the output, the fitted line is $\hat{y} = -87.17 + 1.2765x$.

(b)   For d.f. $= 16$, $t_{0.05} = 1.746$. Therefore, the null hypothesis $H_0 : \beta_1 = 0$ will be rejected at $\alpha = 0.05$ if the observed t value is in the rejection region $R : T \ge 1.746$. According to the output, the observed t is given by $t = \frac{1.2765}{0.2156} = 5.92$, which lies in R. Hence, H_0 is rejected at $\alpha = 0.05$. Furthermore, since the p-value is very small, the data strongly support that $\beta_1 > 0$ which, in turn, indicates that the expected value of weight increases with body length.

(c)   The intercept term is now needed since the p-value is 0.003. The estimated slope has increased.

```
The regression equation is
weight0 = - 87.2 + 1.28 bodylen0


Predictor        Coef      StDev          T      P
Constant       -87.17      28.82      -3.02  0.008
bodylen0       1.2765     0.2156       5.92  0.000


S = 7.980       R-Sq = 68.7%     R-Sq(adj) = 66.7%


Analysis of Variance


Source            DF       SS        MS       F      P
Regression         1     2233.0    2233.0  35.07  0.000
Residual Error    16     1018.8      63.7
Total             17     3251.8


Unusual Observations
Obs bodylen0 weight0    Fit StDev Fit Residual St Resid
  8        146   117.00 99.21       3.31       17.79      2.45R
```

11.65   (a)   From the data, we calculate:

$$\bar{x} = \frac{\sum x}{15} = 16.66 \qquad S_{xx} = \sum (x - \bar{x})^2 = 40.476$$

$$\bar{y} = \frac{\sum y}{15} = 80.04 \qquad S_{xy} = \sum (x - \bar{x})(y - \bar{y}) = 133.804$$

$$S_{yy} = \sum (y - \bar{y})^2 = 629.836$$

So,

$$\hat{\beta}_1 = \frac{S_{xy}}{S_{xx}} = \frac{133.804}{40.476} = 3.306,$$

$$\hat{\beta}_0 = \bar{y} - \hat{\beta}_1 \bar{x} = 80.04 - (3.306)(16.66) = 24.96$$

So, the fitted line is $\hat{y} = 24.96 + 3.306x$.

(b)   The residual sum of squares is SSE = 187.5, and the estimate of σ is

$s = \sqrt{\dfrac{\text{SSE}}{13}} = 3.798$. Since $t_{0.025} = 2.160$ for d.f. = 13, the 95% confidence

interval is given by

$$\hat{\beta}_1 \pm 2.160 \frac{s}{\sqrt{S_{xx}}} = 3.306 \pm 2.160 \left(\frac{3.798}{\sqrt{40.476}} \right) \quad \text{or} \quad (2.02, 4.60).$$

(c)   The predicted temperature ($^{\circ}F$) for $x^* = 15$ is $24.96 + 3.306(15) \approx 74.55$.

11.67 (a) The data for salmon growth are entered into columns C1 and C4 of a Minitab worksheet. And, the data for all salmon growth are stacked into C5 and C6. From the output (seen below), the freshwater growth of salmon is not an effective predictor for its marine growth.

```
STACK (C1 C2) (C3 C4) INTO C5 C6
NAME C5 'FRESHGRW' C6 'MARINGRW'
REGRESS C6 ON 1 PREDICTOR IN C5


THE REGRESSION EQUATION IS
MARINGRW = 470 - 0.258 FRESHGRW
PREDICTOR        COEF        STDEV      T-RATIO         P
CONSTANT        469.62       18.32       25.64      0.000
FRESHGRW        -0.2579      0.1664      -1.55      0.125


S = 40.72        R-SQ = 3.0%       R-SQ(ADJ) = 1.7%



ANALYSIS OF VARIANCE

SOURCE        DF          SS          MS          F          P
REGRESSION     1         3981        3981        2.40      0.125
ERROR         78       129323        1658
TOTAL         79       133305


        CORR C1 C2

CORRELATION OF C1 AND C2 = -0.191
```

(b) The data for male salmon growth are entered into columns C1 and C2. From the output (seen below), the freshwater growth of a male salmon is not an effective predictor of its marine growth.

```
        NAME C1 'FRESHGRW' C2 'MARINGRW'
        REGRESS C2 ON 1 PREDICTOR IN C1

    THE REGRESSION EQUATION IS
    MARINGRW = 478 - 0.236 FRESHGRW

    PREDICTOR        COEF       STDEV      T-RATIO        P
    CONSTANT        478.35      20.30       23.57      0.000
    FRESHGRW       -0.2364      0.1976      -1.20      0.239

    S = 37.05        R-SQ = 3.6%      R-SQ(ADJ) = 1.1%

    ANALYSIS OF VARIANCE

    SOURCE        DF         SS         MS        F        P
    REGRESSION     1        1966       1966      1.43    0.239
    ERROR         38       52168       1373
    TOTAL         39       54134

        CORR C3 C4

    CORRELATION OF C3 AND C4 = 0.040
```

(c)   The data for female salmon growth are entered into columns C3 and C4. From the output (seen on the next page), the freshwater growth of a female salmon is not an effective predictor for its marine growth.

```
        NAME C3 'FRESHGRW' C4 'MARINGRW'
        REGRESS C4 ON 1 PREDICTOR IN C3

    THE REGRESSION EQUATION IS
    MARINGRW = 421 + 0.074 FRESHGRW

    PREDICTOR        COEF       STDEV      T-RATIO        P
    CONSTANT        420.63      35.00       12.02      0.000
    FRESHGRW        0.0742      0.2994       0.25      0.806

    S = 41.55        R-SQ = 0.2%      R-SQ(ADJ) = 0.0%

    ANALYSIS OF VARIANCE

    SOURCE        DF         SS         MS        F        P
    REGRESSION     1         106        106      0.06    0.806
    ERROR         38       65597       1726
    TOTAL         39       65703

        CORR C5 C6

    CORRELATION OF C5 AND C6 = -0.173
```

Chapter 12

REGRESSION ANALYSIS II – MULTIPLE LINEAR REGRESSION AND OTHER TOPICS

12.1 (a) The scatter diagram and fitted line are illustrated below:

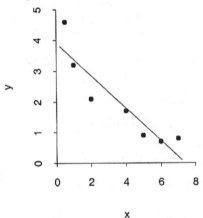

(b) Observe that

$$\bar{x} = \frac{\sum x}{n} = \frac{25.5}{7} = 3.6429 \qquad S_{xx} = \sum (x - \bar{x})^2 = 38.3571$$

$$\bar{y} = \frac{\sum y}{n} = \frac{14.0}{7} = 2.0 \qquad S_{xy} = \sum (x - \bar{x})(y - \bar{y}) = -20.2$$

$$S_{yy} = \sum (y - \bar{y})^2 = 12.64$$

So,

$$\hat{\beta}_1 = \frac{S_{xy}}{S_{xx}} = -0.5266, \quad \hat{\beta}_0 = \bar{y} - \hat{\beta}_1\bar{x} = 2.0 - (-0.5266)(3.6429) = 3.9184$$

The fitted line is $\hat{y} = 3.92 - 0.53x$, which is graphed in part (a).

(c)   Proportion of y variability explained is given by $r^2 = \dfrac{S_{xy}^2}{S_{xx}S_{yy}} = 0.842$.

12.3   (a) $y' = \dfrac{1}{y^{1/3}}$, $x' = x$

        (b) $y' = \dfrac{1}{y}$, $x' = \dfrac{1}{1+x}$

12.5   (a)   The scatter diagram is shown in the below figure (labeled as (i)).
        (b)   The scatter diagram of the transformed data, $x' = \log x$ and $y' = \log y$,
                reveals a more nearly linear relationship – this is illustrated in the figure
                below (labeled as (ii)).

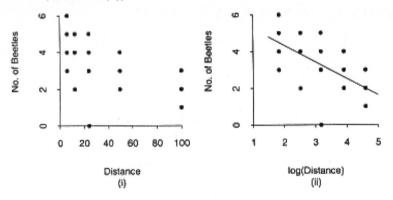

Using Minitab, the original data are entered into columns C1 and C2 – the
output is at the top of the next page.

In order to evaluate the mean and sum of squares for $\log(x)$, we also
calculate the following using Minitab:

```
              STDEV C3
    ST.DEV. =        1.0059
              MEAN C3
    MEAN    =        3.2107
```

```
NAME C1 'DISTANCE' C2 'BEETLES' C3 'LOGEDIST'
LOGE C1 SET IN C3
REGRESS Y IN C2 ON 1 PREDICTOR IN C3


THE REGRESSION EQUATION IS
BEETLES = 6.14 - 0.899 LOGEDIST


PREDICTOR      COEF      STDEV    T-RATIO       P
CONSTANT      6.1e74    0.9918      6.19    0.000
LOGEDIST     -0.8993    0.2954     -3.04    0.007


S = 1.295        R-SQ = 34.0%     R-SQ(ADJ) = 30.3%


ANALYSIS OF VARIANCE


SOURCE        DF        SS         MS        F       P
REGRESSION     1      15.547     15.547     9.27    0.007
ERROR         18      30.203      1.678
TOTAL         19      45.750
```

(c)　Since $(n-1)(\text{Standard Deviation})^2 = 19(1.0059)^2 = 19.225$ and
$t_{0.025} = 2.101$ for d.f. = 18, a 95% confidence interval for β_1 is given by

$$\hat{\beta_1} \pm 2.101 \frac{s}{\sqrt{S_{xx}}} = -0.8993 \pm 2.101(0.2954) \text{ or } (-1.52, -0.28).$$

(d)　At $x = 18$, $\hat{y} = 6.14 - 0.899 \log_e(18) = 3.54$. Since $S_{xx} = 19.225$, a 95% confidence interval is given by

$$3.54 \pm 2.101(1.295)\sqrt{\frac{1}{20} + \frac{(2.8904 - 3.2107)^2}{19.225}} = 3.54 \pm 0.64$$

or (2.90, 4.18).

12.7　When $x_1 = 3$ and $x_2 = -2$, the mean of the response Y is
$$E(Y) = \beta_0 + \beta_1 x_1 + \beta_2 x_2 = -2 - 1(3) + 3(-2) = -11.$$

12.9　(a)　Since $t_{0.025} = 2.110$ for d.f. = 17, the 95% confidence intervals for β_0 and β_2 are
$$\hat{\beta_0} \pm t_{0.025} \times \text{SE}(\hat{\beta_0}) = 3.8 \pm 2.110(1.345) \text{ or } (0.96, 6.64)$$
$$\hat{\beta_2} \pm t_{0.025} \times \text{SE}(\hat{\beta_2}) = -4.2 \pm 2.110(0.4056) \text{ or } (-5.06, -3.34)$$

(b)　Given that $\alpha = 0.05$ and H_1 is right-sided, the rejection region is
$R : T > t_{0.05} = 1.740$ for d.f. = 17. Since the observed value of t is

$$\frac{\hat{\beta}_1 - 6}{SE(\hat{\beta}_1)} = \frac{7.9 - 6}{0.907} = 2.09 \text{, which lies in } R \text{, we reject the null hypothesis}$$

$H_0 : \beta_1 = 6$ in favor of $H_1 : \beta_1 > 6$ at $\alpha = 0.05$.

12.11   (a) $\hat{\beta}_0 = -0.0810, \;\; \hat{\beta}_1 = 0.64588, \;\; \hat{\beta}_2 = 0.8046$

         (b) $\hat{y} = -0.081 + 0.646x_1 + 0.805x_2$. The constant term could be dropped and the model re-fit.

         (c)   The proportion of y variability explained is $R^2 = 0.865$.

         (d) $s^2 = \dfrac{SSE}{n-2} = \text{Error MS} = 0.01023$

12.13   (a)   Given that $\alpha = 0.05$ and H_1 is two-sided, the rejection region is
$R : |T| > t_{0.025} = 2.074$ for d.f. = 22. Since the observed value of t is

$$\frac{\hat{\beta}_1 - 0}{SE(\hat{\beta}_1)} = 7.02 \text{, which lies in } R \text{, we reject the null hypothesis } H_0 : \beta_1 = 0$$

in favor of $H_1 : \beta_1 \neq 0$ at $\alpha = 0.05$.

         (b)   Given that $\alpha = 0.05$ and H_1 is two-sided, the rejection region is
$R : |T| > t_{0.025} = 2.074$ for d.f. = 22. Since the observed value of t for $\hat{\beta}_2$ is
3.29, which lies in R, we reject the null hypothesis $H_0 : \beta_2 = 0$ in favor of
$H_1 : \beta_2 \neq 0$ at $\alpha = 0.05$.

         (c) $\hat{y} = -0.081 + 0.646(1.9) + 0.805(1.0) = 1.951$

         (d)   A 90% confidence interval for β_0, which includes 0, is given by

$$\hat{\beta}_0 \pm t_{0.05} \times SE(\hat{\beta}_0) = -0.081 \pm 2.074(0.1652) \text{ or } (-0.424, 0.262).$$

12.15   (a)   The scatter diagram of y versus $x' = \log_{10} x$ is shown below:

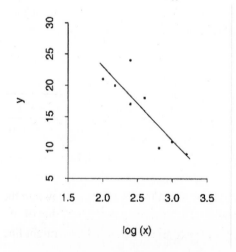

(b) The Minitab output is given below. The fitted line is
$\hat{y} = 46.55 - 11.77 \log_{10} x$ (which is shown in part (a)).

```
The regression equation is
y = 46.6 - 11.8 logt(x)


Predictor     Coef     StDev       T        P
Constant    46.550     7.242     6.43    0.001
logt(x)    -11.772     2.783    -4.23    0.005



S = 3.031        R-Sq - 74.9%


Analysis of Variance


Source          DF       SS       MS       F     P
Regression       1    164.39   164.39   17.90 0.005
Residual Error   6     55.11     9.19
Total            7    219.50
```

(c) From the Minitab output in part (a), $\dfrac{s}{\sqrt{S_{x'x'}}} = 2.783$. You could also

obtain the same result from direct calculation using $S_{x'x'} = 1.1862$. A 90%
confidence interval for β_1 is given by

$$\hat{\beta}_1 \pm 2.101 \frac{s}{\sqrt{S_{x'x'}}} = -11.77 \pm 1.943(3.031) \text{ or } (-17.66, -5.88).$$

(d) At $x = 300$, $\hat{y} = 46.55 - 11.772\left(\log_{10}(300)\right) = 17.39$. Since $\overline{x}' = 2.5739$
and $S_{x'x'} = 1.1862$, a 95% confidence interval for the expected y-value at
$x = 300$ is given by

$$\hat{y} \pm t_{0.025}\, s \sqrt{\frac{1}{n} + \frac{\left((x')^* - \overline{x}'\right)^2}{S_{x'x'}}}$$

$$= 17.39 \pm 2.447(3.031)\sqrt{\frac{1}{8} + \frac{(2.4471 - 2.5739)^2}{1.1862}}$$

$$= 17.39 \pm 2.72$$

or $(14.7, 20.1)$.

12.17 (a) & (b) The scatter diagram of y versus x shown in the figure (part (i))
below reveals a relation along a curve, and that of $y' = \log_{10}(y)$ versus x
shown in the figure (part (ii)) looks like a straight line relation.

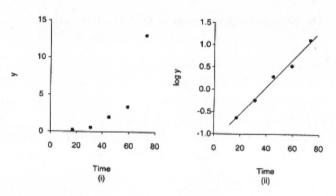

(c)   The fitted line $\log_{10}(y) = -1.16 + 0.0305x$ is shown in part (ii) of the
figure in part (a). The Minitab output is shown below:

```
NAME C1 'TIME' C2 'Y' C3 'LOG(Y)'
LOGT C2 SET C3
REGRESS Y IN C3 ON 1 PREDICTOR IN C1


THE REGRESSION EQUATION IS
LOG(Y) = - 1.16 + 0.0305 TIME


PREDICTOR        COEF      STDEV     T-RATIO         P
CONSTANT      -1.15984    0.09468    -12.25     0.001
TIME          0.030513   0.001926     15.84     0.001


S = 0.08526     R-SQ = 98.8%      R-SQ(ADJ) = 98.4%


ANALYSIS OF VARIANCE


SOURCE      DF      SS       MS       F        P
REGRESSION   1   1.8249   1.8249   251.05    0.001
ERROR        3   0.0218   0.0073
TOTAL        4   1.8467
```

12.19   (a)   A 90% confidence interval for β_1 is given by
$$\hat{\beta}_1 \pm t_{0.05} \times \mathrm{SE}(\hat{\beta}_1) = 1.93 \pm 1.717(0.062) \text{ or } (1.82, 2.04).$$
(b)   Given that $\alpha = 0.05$ and H_1 is left-sided, the rejection region is
$R : T < -t_{0.05} = -1.717$ for d.f. = 22. Since the observed value of t for $\hat{\beta}_2$ is

$$\frac{\hat{\beta}_2 - 25}{SE(\hat{\beta}_2)} = \frac{20.2 - 25}{2.43} = -1.975 \text{, which lies in } R \text{, we reject the null}$$

hypothesis $H_0 : \beta_2 = 25$ in favor of $H_1 : \beta_2 < 25$ at $\alpha = 0.05$.

12.21   (a)   We plot the residual versus the predicted value $\hat{y}$ and the time order, respectively, in parts (i) and (ii) of the below figure.

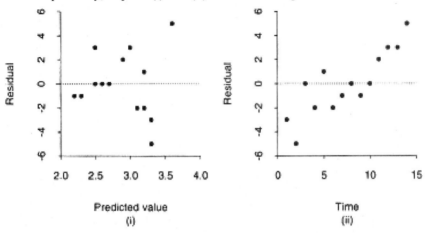

     (b)   The plot (i) does not seem to signify any appreciable violation of the assumptions. The plot (ii) of residual versus time order, however, exhibits a distinct pattern. The residuals tend to steadily increase in time. This indicates a possible violation of the independence assumption.

12.23   Looking at the residuals in time order shown in the figure below, a distinct pattern is apparent. The residuals decrease quite systematically until about the year 15, and then they steadily increase. Residuals adjacent in time have similar magnitude. This pattern casts serious doubt on the independence assumption. Violation of independence is frequent in time series data such as these.

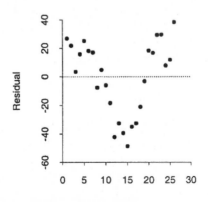

12.25   Use Minitab to find a quadratic fit for these data:

Polynomial Regression

```
Y = 117.920 - 16.0517X + 3.70153X**2
R-Sq = 99.4 %

Analysis of Variance

SOURCE              DF          SS          MS          F           P
Regression           2     43492.7     21746.3     538.007    2.17E-08
Error                7       282.9        40.4
Total                9     43775.6


SOURCE        DF     Seq SS           F          P
Linear         1     41334.4     135.455    2.71E-06
Quadratic      1      2158.3      53.3962   1.62E-04
```

Regression Plot

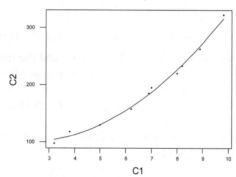

(b)   The proportion of y variability explained is $R^2 = 0.994$, or 99.4%.

(c)   Given that $\alpha = 0.05$ and H_1 is left-sided, the rejection region is
$R : T < -t_{0.05} = -1.895$ for d.f. = 7. Since the observed value of t is

$$\frac{\hat{\beta}_1 - 0}{SE(\hat{\beta}_1)} = \frac{-16.052}{6.356} = -2.525,$$ we reject the null hypothesis $H_0 : \beta_1 = 0$ in

favor of $H_1 : \beta_1 < 0$ at $\alpha = 0.05$.

12.27   (a)   The Minitab output is as follows:

```
SQRT C2 SET IN C3
REGRESS Y IN C3 ON 1 PREDICTOR IN C1

THE REGRESSION EQUATION IS
C3 = - 0.167 + 0.237 C1


PREDICTOR        COEF        STDEV     T-RATIO        P
CONSTANT       -0.1665      0.9323      -0.18      0.863
C1             0.23703      0.02383      9.95      0.000


S = 0.9563      R-SQ = 92.5%      R-SQ(ADJ) = 91.6%


ANALYSIS OF VARIANCE

SOURCE         DF          SS          MS         F         P
REGRESSION      1        90.456      90.456     98.91     0.000
ERROR           8         7.316       0.915
TOTAL           9        97.773
```

From the output, note that the fitted line is $\hat{y}' = -0.167 + 0.237x$ and $r^2 = 0.925$. The constant term could be dropped and the model re-fit.

   (b)   Since $t_{0.025} = 2.306$ for d.f. = 8, a 95% confidence interval for β_1 is given

by $\hat{\beta}_1 \pm 2.306 \dfrac{S}{\sqrt{S_{xx}}} = 0.23703 \pm 2.306(0.02383)$ or $(0.182, 0.292)$.

   (c)   Since $\bar{x} = 37$ and $S_{xx} = 1610$, a 95% confidence interval (for $x = 45$) is given by

$$\left[-0.167 + 0.237(45) \right] \pm 2.306(0.9563) \sqrt{\frac{1}{10} + \frac{(45-37)^2}{1610}} \text{ or } (9.67, 11.32).$$

12.29   (a) $\hat{y} = 50.4 + 0.1907x_2$ and $r^2 = 0.03$. The Minitab output is shown on the next page:

```
        NAME C1 'X1' C2 'X2' C3 'X3' C4 'Y'
        REGRESS Y IN C4 ON 1 PREDICTOR IN C2

THE REGRESSION EQUATION IS
Y = 50.4 + 0.191 X2

PREDICTOR    COEF    STDEV   T-RATIO        P
CONSTANT     50.40   17.57     2.87    0.010
X2           0.1907  0.2561    0.74    0.466

S = 19.95         R-SQ = 3.0%  R-SQ(ADJ) = 0.0%

ANALYSIS OF VARIANCE

SOURCE       DF     SS      MS      F     P
REGRESSION   1    220.8   220.8   0.55  0.466
ERROR        18  7164.2   398.0
TOTAL        19  7385.0
```

(b) $\hat{y} = -92.32 + 0.583x_1 - 0.1494x_2 + 35.07x_3$ and $R^2 = 0.586$. The Minitab output is shown below:

```
        REGRESS Y IN C4 ON 3 PREDICTORS IN C1 C2 C3

   THE REGRESSION EQUATION IS
   Y = - 92.3 + 0.583 X1 - 0.149 X2 + 35.1 X3

   PREDICTOR    COEF    STDEV    T-RATIO        P
   CONSTANT    -92.32   33.22     -2.78    0.013
   X1           0.5830  0.3206     1.82    0.088
   X2          -0.1494  0.1923    -0.78    0.449
   X3          35.07    11.25      3.12    0.007

   S = 13.83    R-SQ = 58.6%    R-SQ(ADJ) = 50.8%

   ANALYSIS OF VARIANCE

   SOURCE       DF      SS      MS      F     P
   REGRESSION   3    4325.7  1441.9   7.54  0.002
   ERROR        16   3059.2   191.2
   TOTAL        19   7385.0
   SOURCE       DF     SEQ SS
   X1           1      2466.6
   X2           1         1.4
   X3           1      1857.8
```

(c)   Even three variables do not predict well. In fact, the GPA (x_3) could predict almost as well by itself. We summarize the results in the following table:

Predictor	x_3	x_3 and x_1	x_3, x_1, and x_2
R^2	0.495	0.570	0.586

12.31   The design matrix X of the model $Y = \beta_0 + \beta_1 x_1 + \beta_2 x_2 + e$ is

$$X = \begin{bmatrix} 1 & 1 & 14 \\ 1 & 2 & 44 \\ 1 & 2 & 20 \\ 1 & 4 & 36 \\ 1 & 4 & 66 \\ 1 & 5 & 59 \\ 1 & 7 & 100 \\ 1 & 7 & 95 \\ 1 & 8 & 38 \end{bmatrix}$$

12.33   Using Minitab, the data for gender, initial and final number of sit-ups are entered into columns C1, C8, and C9, respectively. The fitted equation is

final number $= 9.999 + 0.155$ (gender) $+ 0.9015$ (initial number)

With the high p-value of 0.899 shown in the output (below), we cannot reject the hypothesis that the coefficient of predictor 'gender' is zero. In fact, a simple linear fit with initial number of sit-ups as predictor has $r^2 = 0.715$.

```
NAME C1 'GENDER' C8 'PRETEST' C9 'POSTTEST'
REGRESS Y IN C9 ON 2 PREDICTORS IN C1 AND C8

THE REGRESSION EQUATION IS
POSTTEST = 10.0 + 0.15 GENDER + 0.902 PRETEST

PREDICTOR        COEF        STDEV      T-RATIO        P
CONSTANT         9.999       3.640        2.75      0.007
GENDER           0.155       1.211        0.13      0.899
PRETEST          0.90150     0.06710     13.43      0.000

S = 5.211        R-SQ = 71.5%     R-SQ(ADJ) = 70.8%

ANALYSIS OF VARIANCE

SOURCE        DF      SS        MS         F         P
REGRESSION     2    5313.4    2656.7     97.82     0.000
ERROR         78    2118.4      27.2
TOTAL         80    7431.8

SOURCE        DF    SEQ SS
GENDER         1     411.6
PRETEST        1    4901.7
```

Chapter 13

ANALYSIS OF CATEGORICAL DATA

13.1 The null hypothesis is $H_0 : p_1 = p_2 = ... = p_6 = \frac{1}{6}$ and the alternative hypothesis H_1 is that at least two proportions are different. Under H_0, the expected frequency of each cell is $320\left(\frac{1}{6}\right)$. The χ^2 statistic for goodness-of-fit is calculated as follows:

Face Number	1	2	3	4	5	6	Total
Observed Frequency (O)	39	63	56	67	57	38	320
Expected Frequency (E)	$\frac{320}{6}$	$\frac{320}{6}$	$\frac{320}{6}$	$\frac{320}{6}$	$\frac{320}{6}$	$\frac{320}{6}$	320
$\frac{(O-E)^2}{E}$	3.852	1.752	0.133	3.502	0.252	4.408	$13.90 = \chi^2$

We take $\alpha = 0.05$. For d.f. = 5, we find that $\chi^2_{0.05} = 11.07$, so that the rejection region is $R : \chi^2 \geq 11.07$. Since the observed value of $\chi^2 = 13.90$ lies in R, we reject H_0 at $\alpha = 0.05$. As such, the model of a fair die is contradicted.

13.3 The null hypothesis is $H_0 : p_1 = 0.4$, $p_2 = 0.4$, $p_3 = 0.1$, $p_4 = 0.1$ and the alternative hypothesis H_1 is that at least one of these proportions is not correct. Under H_0, multiplying these probabilities by $n = 100$, the expected frequencies are found to be 40, 40, 10, and 10. The χ^2 statistic for goodness-of-fit is calculated as follows:

Blood Type	O	A	B	AB	Total
Observed Frequency (O)	40	44	10	6	100
Expected Frequency (E)	40	40	10	10	100
$\dfrac{(O-E)^2}{E}$	0.00	0.40	0.00	1.60	$2.00 = \chi^2$

We take $\alpha = 0.05$. For d.f. = 3, we find that $\chi^2_{0.05} = 7.81$, so that the rejection region is $R : \chi^2 \geq 7.81$. Since the observed value of $\chi^2 = 2.00$ does not lie in R, we do not reject H_0 at $\alpha = 0.05$.

13.5   Let p_1, p_2, p_3, and p_4 denote the population proportions of walnuts, hazelnuts, almonds, and pistachios, respectively. The null hypothesis is $H_0 : p_1 = 0.45$, $p_2 = 0.20$, $p_3 = 0.20$, $p_4 = 0.15$ and the alternative hypothesis H_1 is that at least one of these proportions is not correct. Under H_0, multiplying these probabilities by $n = 240$, the expected frequencies are found to be 108, 48, 48, and 36. The χ^2 statistic for goodness-of-fit is calculated as follows:

Nut Type	Walnuts	Hazelnuts	Almonds	Pistachios	Total
Observed Frequency (O)	95	70	33	42	240
Expected Frequency (E)	108	48	48	36	240
$\dfrac{(O-E)^2}{E}$	1.565	10.083	4.688	1.000	$17.336 = \chi^2$

We take $\alpha = 0.025$. For d.f. = 3, we find that $\chi^2_{0.025} = 9.35$, so that the rejection region is $R : \chi^2 \geq 9.35$. Since the observed value of $\chi^2 = 17.336$ lies in R, we reject H_0 at $\alpha = 0.025$. We conclude that there is strong evidence of mislabeling.

13.7   For the particular geographical region, let p_1, p_2, ..., p_6 denote the population proportions of accidental deaths due to motor vehicle, falls, drowning, burns, poison, and other reasons, respectively. The null hypothesis is $H_0 : p_1 = 0.46$, $p_2 = 0.15$, $p_3 = 0.04$, $p_4 = 0.04$, $p_5 = 0.08$, $p_6 = 0.23$ and the alternative hypothesis H_1 is that at least one of these proportions is not correct. Under H_0, multiplying these probabilities by $n = 932$, the expected frequencies are found to be 428.72, 139.80, 37.28, 37.28, 74.56, and 214.36. The χ^2 statistic for goodness-of-fit is calculated as follows:

Reason	Motor Vehicle	Falls	Drowning	Burns	Poison	Other	Total
Observed Frequency (O)	462	171	76	57	92	74	932
Expected Frequency (E)	428.72	139.80	37.28	37.28	74.56	214.36	932
$\dfrac{(O-E)^2}{E}$	2.583	6.963	40.216	10.431	4.079	91.906	$\begin{array}{c}156.18\\ = \chi^2\end{array}$

We take $\alpha = 0.01$. For d.f. = 5, we find that $\chi^2_{0.01} = 15.09$, so that the rejection region is $R: \chi^2 \geq 15.09$. Since the observed value of $\chi^2 = 156.18$ lies in R, we reject H_0 at $\alpha = 0.01$. As such, we conclude that the pattern is significantly different – the differences in categories 3, 4 and 6 are conspicuous from the large values of $\dfrac{(O-E)^2}{E}$.

13.9 (a) From the binomial table for $n = 3$ and $p = 0.4$, we find that
$$P[X = 0] = 0.216$$
$$P[X = 1] = 0.648 - 0.216 = 0.432$$
$$P[X = 2] = 0.936 - 0.648 = 0.288$$
$$P[X = 3] = 1.000 - 0.936 = 0.064$$

(b) Let p_0, p_1, p_2, and p_3 denote the probabilities of the four categories: 0, 1, 2 and 3 males in litter, respectively. The null hypothesis is $H_0: p_0 = 0.216$, $p_1 = 0.432$, $p_2 = 0.288$, $p_3 = 0.064$ and the alternative hypothesis H_1 is that at least one of these proportions is not correct. Under H_0, multiplying these probabilities by $n = 80$, the expected frequencies are found to be 17.28, 34.56, 23.04, and 5.12. The χ^2 statistic for goodness-of-fit is calculated as follows:

Number of Males	0	1	2	3	Total
Observed Frequency (O)	19	32	22	7	80
Expected Frequency (E)	17.28	34.56	23.04	5.12	80
$\dfrac{(O-E)^2}{E}$	0.171	0.190	0.047	0.690	1.098 $= \chi^2$

We take $\alpha = 0.05$. For d.f. = 2, we find that $\chi^2_{0.05} = 5.99$, so that the rejection region is $R : \chi^2 \geq 5.99$. Since the observed value of $\chi^2 = 1.098$ does not lie in R, we do not reject H_0 at $\alpha = 0.05$. As such, we conclude that the binomial model is not contradicted.

13.11 (a) We summarize the information in the table below:

	Open all mail	Don't open all mail	Total
Males	414	386	800
Females	532	368	900
Total	946	754	1700

(b) Let p_1 and p_2 be the probabilities that a person opens all of his/her mail for males and females, respectively.
The null hypothesis of homogeneity is $H_0 : p_1 = p_2$.

(c) We take $\alpha = 0.05$. For d.f. = 1, we have $\chi^2_{0.05} = 3.84$, so the rejection region is $R : \chi^2 \geq 3.84$. The χ^2 statistic for homogeneity is calculated as follows:

Expected Values:

	Open all mail	Don't open all mail	Total
Males	445.176	354.824	800
Females	500.823	399.176	900
Total	946	754	1700

$\dfrac{(O-E)^2}{E}$:

	Open all mail	Don't open all mail
Males	2.183	2.739
Females	1.941	2.435

Then, $\chi^2 = 2.183 + 2.739 + 1.941 + 2.435 = 9.298$, which lies in R. So, we reject H_0 (of equal probabilities of opening all mail) at $\alpha = 0.05$. Furthermore, the p-value is about 0.02.

(d)   The largest contribution to χ^2 comes from the mail – don't open cell, where the observed count is higher than expected. The female – don't open count is lower than expected.

13.13   Denote by p_{A1}, p_{A2}, p_{A3}, and p_{A4} the probabilities of response in the categories 'none', 'slight', 'moderate', and 'severe', respectively, under the use of Brand A pills. Similarly, use p_{B1}, p_{B2}, p_{B3}, and p_{B4} for Brand B. We are to test the null hypothesis of homogeneity $H_0 : p_{Aj} = p_{Bj}$ $(j=1,2,3,4)$.

The χ^2 statistic for homogeneity is calculated as follows:

```
Expected counts are printed below observed counts

          C1       C2       C3       C4     Total
    1     18       17        6        4       45
          14.50    15.50    10.00     5.00

    2     11       14       14        6       45
          14.50    15.50    10.00     5.00

  Total      29       31       20       10       90

ChiSq =  0.845 +  0.145 +  1.600 +  0.200 +
         0.845 +  0.145 +  1.600 +  0.200 = 5.580
df = 3
```

We take $\alpha = 0.05$. For d.f. = 3, we find that $\chi^2_{0.05} = 7.81$, so that the rejection region is $R: \chi^2 \geq 7.81$. Since the observed $\chi^2 = 5.580$ (from the Minitab output) does not lie in R, we do not reject H_0 at $\alpha = 0.05$. As such, we fail to conclude that the two pills are significantly different in quality.

13.15 (a) Denote by p_1, p_2, p_3 and p_4 the probabilities that a paper will be taken from sites 1, 2, 3, and 4, respectively. We are to test the null hypothesis of homogeneity $H_0 : p_1 = p_2 = p_3 = p_4$.

(b) First, we calculate the number of papers taken from each site. For site 1, the number taken is $50 - 17 = 33$; the others are computed similarly. The χ^2 statistic for homogeneity is calculated using Minitab – the output is shown at the top of the next page.

We take $\alpha = 0.05$. For d.f. = 3, we find that $\chi^2_{0.05} = 7.81$, so that the rejection region is $R: \chi^2 \geq 7.81$. Since the observed $\chi^2 = 9.780$ (from the Minitab output) lies in R, we reject H_0 at $\alpha = 0.05$. As such, we conclude that the proportions of papers taken differ between the sites.

(c) A 95% confidence interval for a population proportion p is given by

$$\hat{p} \pm 1.96 \sqrt{\frac{\hat{p}\hat{q}}{n}} .$$

We calculate such an interval for each of the four proportions, as follows:

```
Chi-Square Test


Expected counts are printed below observed counts

           Remain    Taken    Total
    1        17        33        50
            14.62     35.38

    2        12        35        47
            13.74     33.26

    3         7        41        48
            14.03     33.97

    4        21        29        50
            14.62     35.38

Total        57       138       195

    Chi-Sq =  0.389 +  0.161 +
              0.220 +  0.091 +
              3.523 +  1.455 +
              2.789 +  1.152 = 9.780
    DF = 3, P-Value = 0.021
```

For $\hat{p}_1$: $\hat{p}_1 = \frac{33}{50} = 0.66$, $0.66 \pm 1.96 \sqrt{\frac{(0.66)(0.34)}{50}} = 0.66 \pm 0.13$

or $(0.53, 0.79)$

For $\hat{p}_2$: $\hat{p}_2 = \frac{35}{47} = 0.745$, $0.745 \pm 1.96 \sqrt{\frac{(0.745)(0.255)}{47}} = 0.745 \pm 0.125$

or $(0.62, 0.87)$

For $\hat{p}_3$: $\hat{p}_3 = \frac{41}{48} = 0.854$, $0.854 \pm 1.96 \sqrt{\frac{(0.854)(0.156)}{48}} = 0.854 \pm 0.099$

or $(0.754, 0.953)$

For $\hat{p}_4$: $\hat{p}_4 = \frac{29}{50} = 0.58$, $0.58 \pm 1.96 \sqrt{\frac{(0.58)(0.42)}{50}} = 0.58 \pm 0.137$

or $(0.44, 0.72)$

13.17 (a) We are to test $H_0 : p_3 = p_4$ versus $H_1 : p_3 > p_4$.

Since H_1 is one-sided, the χ^2 test is not appropriate. As such, we use the test

statistic $Z = \dfrac{\hat{p}_3 - \hat{p}_4}{\sqrt{\hat{p}\hat{q}}\sqrt{\dfrac{1}{n_1} + \dfrac{1}{n_2}}}$

Since H_1 is right-side, the rejection region is of the form $R : Z \geq c$. We have the following:

$$\hat{p}_3 = \tfrac{41}{48} = 0.854, \qquad \hat{p}_2 = \tfrac{29}{50} = 0.580$$

$$\text{Pooled estimate } \hat{p} = \frac{41 + 29}{48 + 50} = 0.714.$$

The value of the observed z is $\dfrac{0.854 - 0.580}{\sqrt{(0.714)(0.286)}\sqrt{\dfrac{1}{48} + \dfrac{1}{50}}} = 3.00$. The

associated p-value is $P[Z \geq 3.00] = 0.0013$. This small p-value suggests there is strong evidence in support of H_1.

(b) Observe that $\hat{p}_3 - \hat{p}_4 = 0.854 - 0.580 = 0.274$.

$$\text{Estimated S.E.} = \sqrt{\frac{\hat{p}_3\hat{q}_3}{n_3} + \frac{\hat{p}_4\hat{q}_4}{n_4}} = \sqrt{\frac{(0.854)(0.146)}{48} + \frac{(0.580)(0.420)}{50}} = 0.0864$$

So, a 95% confidence interval for $p_3 - p_4$ is given by

$$\left(\hat{p}_3 - \hat{p}_4\right) \pm z_{0.05/2}\sqrt{\frac{\hat{p}_3\hat{q}_3}{n_3} + \frac{\hat{p}_4\hat{q}_4}{n_4}} = 0.274 \pm 1.96(0.0864) = 0.274 \pm 0.169$$

or $(0.11, 0.44)$.

13.19 (a) The calculations are identical.

(b) The Minitab output is as follows:

```
Chi-Square Test


Expected counts are printed below observed counts

           Open Dont Ope    Total
    1       414      386      800
           445.18   354.82

    2       532      368      900
           500.82   399.18

  Total     946      754     1700

  Chi-Sq =   2.183 +   2.739 +
             1.941 +   2.435 = 9.298
  DF = 1, P-Value = 0.002
```

We take $\alpha = 0.05$. For d.f. = 1, we find that $\chi^2_{0.05} = 3.84$, so that the rejection region is $R : \chi^2 \ge 3.84$. Since the observed $\chi^2 = 9.298$ (from the Minitab output) lies in R, we reject H_0 at $\alpha = 0.05$. Furthermore, the associated p-value is 0.002. As such, the proportions of males and females who open all of their mail are significantly different.

(c)   The Minitab output is as follows:

```
Expected counts are printed below observed counts

        C1      C2       C3     Total
        1       38       15       7        60
                22.50   23.10   14.40

        2       22       32       16       70
                26.25   26.95   16.80

        3       15       30       25       70
                26.25   26.95   16.80

Total           75       77       48       200
        ChiSq = 10.678 +  2.840 +  3.803 +
                 0.688 +  0.946 +  0.038 +
                 4.821 +  0.345 +  4.002 = 28.162
     df = 4
```

We take $\alpha = 0.01$. For d.f. = 4, we find that $\chi^2_{0.01} = 13.28$, so that the rejection region is $R : \chi^2 \ge 13.28$. Since the observed $\chi^2 = 28.162$ (from the Minitab output) lies in R, we reject H_0 at $\alpha = 0.01$. Comparing observed and expected frequencies, we see that the bone loss is higher in the control group than the activity group.

13.21   We test the null hypothesis that the pattern of appeals decision and the type of representation are independent. The χ^2 statistic is calculated using Minitab as follows:

```
Expected counts are printed below observed counts

        C1      C2       C3     Total
        1       59       108      17       184
                74.18   98.32   11.50

        2       70       63       3        136
                54.83   72.68   8.50

Total           129      171      20       320

        ChiSq =  3.105 +  0.952 +  2.630 +
                 4.200 +  1.288 +  3.559 = 15.734
     df = 2
```

We take $\alpha = 0.05$. For d.f. = 2, we find that $\chi^2_{0.05} = 5.99$, so that the rejection region is $R : \chi^2 \geq 5.99$. Since the observed $\chi^2 = 15.734$ (from the Minitab output) lies in R, we reject H_0 at $\alpha = 0.05$. We conclude that the patterns of appeals decision are significantly different between the two types of representation.

13.23 We test the null hypothesis of independence between union membership and attitude toward spending on social welfare. The χ^2 statistic is calculated using Minitab as follows:

```
Expected counts are printed below observed counts

              C1         C2         C3     Total
    1        112         36         28       176
            86.24      45.76      44.00

    2         84         68         72       224

           109.76      58.24      56.00

  Total      196        104        100       400

  ChiSq =  7.695 +   2.082 +   5.818 +
           6.046 +   1.636 +   4.571 = 27.847

  df = 2
```

We take $\alpha = 0.01$. For d.f. = 2, we find that $\chi^2_{0.01} = 9.21$, so that the rejection region is $R : \chi^2 \geq 9.21$. Since the observed $\chi^2 = 27.847$ (from the Minitab output) lies in R, we reject H_0 at $\alpha = 0.01$. We conclude that attitudes and union membership are dependent. There are significant differences between the attitudes of the union and non-union groups.

13.25 The null hypothesis is that group and stopping response are independent. The cell probabilities are the product of the marginal probabilities. The χ^2 statistic is calculated using Minitab as follows:

```
        Expected counts are printed below observed counts

               C1       C2    Total
        1       9        9       18
               6.55    11.45

        2       8       12       20
               7.27    12.73


        3       3       14       17
               6.18    10.82


  Total        20       35       55


    ChiSq =  0.920 +  0.526 +
             0.073 +  0.042 +
             1.638 +  0.936 = 4.134
  df = 2
```

We take $\alpha = 0.05$. For d.f. $= 2$, we find that $\chi^2_{0.05} = 5.99$, so that the rejection region is $R : \chi^2 \geq 5.99$. Since the observed $\chi^2 = 4.1334$ (from the Minitab output) does not lie in R, we do not reject H_0 at $\alpha = 0.05$. We conclude that, under independence, the groups are not significantly different in their response.

13.27 (a) They are identical except for the third decimal in the Chi-sq entry.
       (b) The Minitab output is as follows:

```
      Expected counts are printed below observed counts

                    Yes         No      Not Sure    Total
      Democrat      138         83         64         285
                    115.14     85.50      84.36

      Republican    64          67         84         215
                    86.68      64.50      63.64

      Total         202        150        148         500

  Chi-Sq   =   4.539 + 0.073 + 4.914 +
               6.016 + 0.097 + 6.514 = 22.153

  DF = 2    p-value 0.000
```

Since the p-value is so small, the null hypothesis would be rejected for any reasonable value of α This provides strong evidence against independence.

13.29 Denote by p_0, p_1, ... , p_9 the probabilities of the integers 0, 1, 2, ... , 9, respectively. The null hypothesis that all ten integers are equally likely is formalized by $H_0 : p_0 = p_1 = ... = p_9 = \frac{1}{10}$ and the alternative hypothesis H_1 is that at least two of these proportions are different. Under H_0, the expected frequency of each cell is $500\left(\frac{1}{10}\right) = 50$. The χ^2 statistic for goodness-of-fit is calculated as follows:

Integer	0	1	2	3	4	5	6	7	8	9	Total
Observed Frequency (O)	41	58	51	61	39	56	45	35	62	52	500
Expected Frequency (E)	50	50	50	50	50	50	50	50	50	50	500
$\dfrac{(O-E)^2}{E}$	1.62	1.28	0.02	2.42	2.42	0.72	0.50	4.50	2.88	0.08	$16.44 = \chi^2$

We take $\alpha = 0.05$. For d.f. = 9, we find that $\chi^2_{0.05} = 16.92$, so that the rejection region is $R : \chi^2 \geq 16.92$. Since the observed value of $\chi^2 = 16.44$ does not lie in R, we do not reject H_0 at $\alpha = 0.05$. As such, the data do not demonstrate any bias.

13.31 Denote by p_1, p_2, p_3, and p_4 the probabilities of birth in the four consecutive quarters. The null hypothesis is $H_0 : p_1 = 0.4$, $p_2 = 0.2$, $p_3 = 0.2$, $p_4 = 0.2$ and the alternative hypothesis H_1 is that at least one of these proportions is not correct. Under H_0, multiplying these probabilities by $n = 300$, the expected frequencies are found to be 120, 60, 60, and 60. The χ^2 statistic for goodness-of-fit is calculated as follows:

Quarter	Jan – Mar	Apr – Jun	Jul – Sep	Oct – Dec	Total
Observed Frequency (O)	55	29	26	41	151
Expected Frequency (E)	60.4	30.2	30.2	30.2	151
$\dfrac{(O-E)^2}{E}$	0.483	0.048	0.584	3.862	$4.977 = \chi^2$

We take $\alpha = 0.10$. For d.f. = 3, we find that $\chi^2_{0.10} = 6.25$, so that the rejection region is $R : \chi^2 \geq 6.25$. Since the observed value of $\chi^2 = 4.977$ does not lie in R, we do not reject H_0 at $\alpha = 0.05$. As such, the stated conjecture is not contradicted.

13.33   Denote by p_1 and p_2 the population proportion of persons having hepatitis in the two groups 'vaccinated' and 'not vaccinated', respectively.   We are to test $H_0 : p_1 = p_2$ versus $H_1 : p_1 \neq p_2$.   The χ^2 statistic for homogeneity is calculated using Minitab as follows:

```
Expected counts are printed below observed counts

              C1       C2     Total
     1        11      538       549
            41.06   507.94

     2        70      464       534
            39.94   494.06

Total        81     1002      1083

   ChiSq = 22.008 +  1.779 +

                22.626 +   1.829 = 48.242
       df = 1
```

We take $\alpha = 0.05$.   For d.f. = 1, we find that $\chi^2_{0.05} = 3.84$, so that the rejection region is $R : \chi^2 \geq 3.84$.   Since the observed value of $\chi^2 = 48.242$ lies in R, we reject H_0 at $\alpha = 0.05$.   This is very strong evidence in support of H_1.

13.35   (a)   The two response categories are 'free of pain' and 'not free of pain'.   The frequencies of the latter category are obtained by subtracting those of the first from the corresponding 'number of patients assigned'.   The 4×2 contingency table is presented here, along with the calculations (from Minitab) for the χ^2 statistic.

```
      Expected counts are printed below observed counts

            Free    Not free Total
     1       23        30       53
            27.45     25.55

     2       30        17       47

            24.34     22.66

     3       19        32       51
            26.42     24.58

     4       29        15       44
            22.79     21.21

  Total      101       94      195

  Chi-Sq =  0.722 +  0.776 +
            1.314 +  1.412 +
            2.082 +  2.237 +
            1.692 +  1.818 = 12.053
  DF = 3, P-Value = 0.007
```

We take $\alpha = 0.05$. For d.f. = 3, we find that $\chi^2_{0.05} = 7.81$, so that the rejection region is $R : \chi^2 \geq 7.81$. Since the observed value of $\chi^2 = 12.053$ lies in R, we reject H_0 at $\alpha = 0.05$. We conclude that there are significant differences in the effectiveness of the four drugs.

(b)   A 95% confidence interval for a population proportion p is given by

$$\hat{p} \pm 1.96 \sqrt{\frac{\hat{p}\hat{q}}{n}}.$$

We calculate such an interval for each of the four proportions, as follows:

For Drug 1: $\hat{p}_1 = \frac{23}{53} = 0.434$, $0.442 \pm 1.645 \sqrt{\frac{(0.434)(0.566)}{53}}$

$= 0.434 \pm 0.112$ or $(0.32, 0.55)$

For Drug 2: $\hat{p}_2 = \frac{30}{47} = 0.638$, $0.625 \pm 1.645 \sqrt{\frac{(0.638)(0.362)}{47}}$

$= 0.638 \pm 0.115$ or $(0.52, 0.75)$

For Drug 3: $\hat{p}_3 = \frac{19}{51} = 0.373$, $0.373 \pm 1.645 \sqrt{\frac{(0.373)(0.627)}{51}}$

$= 0.373 \pm 0.111$ or $(0.26, 0.48)$

For Drug 4: $\hat{p}_4 = \frac{29}{44} = 0.659, \quad 0.659 \pm 1.645 \sqrt{\dfrac{(0.659)(0.341)}{44}}$

$$= 0.659 \pm 0.118 \quad \text{or} \quad (0.54, 0.78)$$

13.37 (a) We are to test $H_0 : p_4 = p_3$ versus $H_1 : p_4 > p_3$.

Since H_1 is one-sided, the χ^2 test is not appropriate. As such, we use the test statistic

$$Z = \frac{\hat{p}_4 - \hat{p}_3}{\sqrt{\hat{p}\hat{q}} \sqrt{\dfrac{1}{n_1} + \dfrac{1}{n_2}}}$$

Since H_1 is right-side, the rejection region is of the form $R : Z \geq c$. We have the following:

$$\hat{p}_3 = \frac{19}{51} = 0.373, \qquad \hat{p}_4 = \frac{29}{44} = 0.659$$

$$\text{Pooled estimate } \hat{p} = \frac{29 + 19}{44 + 51} = 0.505.$$

The value of the observed z is $\dfrac{0.659 - 0.373}{\sqrt{(0.505)(0.495)} \sqrt{\dfrac{1}{44} + \dfrac{1}{51}}} = 2.78$. The

associated p-value is $P[Z \geq 2.78] = 0.0027$. This small p-value suggests there is strong evidence in support of H_1.

(b) Observe that $\hat{p}_4 - \hat{p}_3 = 0.659 - 0.373 = 0.286$.

$$\text{Estimated S.E.} = \sqrt{\frac{\hat{p}_3 \hat{q}_3}{n_3} + \frac{\hat{p}_4 \hat{q}_4}{n_4}} = \sqrt{\frac{(0.373)(0.627)}{51} + \frac{(0.659)(0.341)}{44}} = 0.0985$$

So, a 95% confidence interval for $p_4 - p_3$ is given by

$$\left(\hat{p}_4 - \hat{p}_3 \right) \pm z_{0.05/2} \sqrt{\frac{\hat{p}_3 \hat{q}_3}{n_3} + \frac{\hat{p}_4 \hat{q}_4}{n_4}} = 0.286 \pm 1.96(0.0985) = 0.286 \pm 0.193$$

or $(0.09, 0.48)$.

13.39 We test the null hypothesis that the duration of marriage is independent of the period of acquaintance before marriage. The χ^2 statistic is calculated (using Minitab) as follows:

```
Expected counts are printed below observed counts
        C1        C2     Total
  1     11        8       19
        10.27     8.73


  2     28        24      52
        28.11     23.89


  3     21        19      40
        21.62     18.38


Total   60        51      111

  ChiSq =  0.052 +  0.061 +
           0.000 +  0.000 +
           0.018 +  0.021 = 0.153
df = 2
```

For $\alpha = 0.05$, the tabulated value is $\chi^2_{0.05} = 5.99$. The observed χ^2 is not significant. As such, the null hypothesis of independence between period of acquaintanceship and duration of marriage is not contradicted.

13.41 We test the null hypothesis of independence. The χ^2 statistic is calculated (using Minitab) as follows:

```
Expected counts are printed below observed counts

        One   Neither   Total
Innoc   27      20       47
        22.43   24.57

Unbal   36      49       85
        40.57   44.43

Total   63      69       132


  Chi-Sq =  0.930 +  0.849 +
            0.514 +  0.470 = 2.764
DF = 1, P-Value = 0.096
```

We take $\alpha = 0.05$. For d.f. = 1, we find that $\chi^2_{0.05} = 3.84$, so that the rejection region is $R: \chi^2 \geq 3.84$. Since the observed value of $\chi^2 = 2.764$ does not lie in R, we do not reject H_0 at $\alpha = 0.05$.

13.43 (a)   In Table 14, the expected frequency of the first cell is $75\left(\frac{100}{300}\right) = 25$, which is the same as the observed frequency. In the same manner, it is seen that the expected and observed frequencies are identical in every cell. Consequently, $\chi^2 = 0$. This is also the case for Table 15.

(b)   The χ^2 statistic is calculated (using Minitab) for the pooled data in Table 16 as follows:

```
Expected counts are printed below observed counts

             Offer   Denied    Total
    Male      175      100       275
             148.96   126.04

  Female      150      175       325
             176.04   148.96

  Total       325      275       600

     ChiSq =  4.553 +  5.381 +
              3.852 +  4.553 = 18.338
  df = 1
```

For d.f. = 1, we find that $\chi^2_{0.05} = 3.84$, so the null hypothesis is rejected at $\alpha = 0.05$.

(c) Note that for secretarial positions the proportion of candidates receiving an offer is $25/75 = 1/3$ for males and $75/225 = 1/3$ for females. The proportions for sales positions are $150/200 = 3/4$ for males and $75/100 = 3/4$ for females. Although the rates are equal within each table, we see that in the pooled table, the rates are $175/272 = 0.64$ for males and $150/325 = 0.46$ for females. This discrepancy arises because in Table 14, $75/300$ (or 25%) are male applicants, whereas in Table 15 there are 67%. Thus, pooling the tables results in a rather uneven mix. Since the overall rates of offer are very different for the two kinds of positions, the tables should not be combined.

Chapter 14

ANALYSIS OF VARIANCE (ANOVA)

14.1 (a) We first find $k = 4$, $\bar{y} = 6$, $\bar{y}_1 = 1$, $\bar{y}_2 = 0$, $\bar{y}_3 = -3$, and $\bar{y}_4 = 2$. Thus,

$$
\begin{array}{cccc}
\text{Obs.} & \text{Grand mean} & \text{Tr. Effect} & \text{Residuals} \\
y_{ij} & \bar{y} & \bar{y}_i - \bar{y} & y_{ij} - \bar{y}_i
\end{array}
$$

$$
\begin{bmatrix} 5 & 9 \\ 8 & 4 \\ 4 & 2 \\ 7 & 9 \end{bmatrix} = \begin{bmatrix} 6 & 6 \\ 6 & 6 \\ 6 & 6 \\ 6 & 6 \end{bmatrix} + \begin{bmatrix} 1 & 1 \\ 0 & 0 \\ -3 & -3 \\ 2 & 2 \end{bmatrix} + \begin{bmatrix} -2 & 2 \\ 2 & -2 \\ 1 & -1 \\ -1 & 1 \end{bmatrix}
$$

(b) Treatment SS $= 2(1)^2 + 2(0)^2 + 2(-3)^2 + 2(2)^2 = 28$
    Residual SS $= (-2)^2 + (2)^2 + \ldots + (-1)^2 + 1^2 = 20$
    Mean SS $= 8(6)^2 = 288$
    Total SS $= 5^2 + 9^2 + \ldots + 9^2 = 336$
    Total SS (corrected) $= (5 - 6)^2 + \ldots + (9 - 6)^2 = 336 - 288 = 48$

(c) Error d.f. $= \sum n_i - k = 2 + 2 + 2 + 2 - 4 = 4$
    Treatment d.f. $= k - 1 = 4 - 1 = 3$

(d) The analysis-of-variance table is

ANOVA Table

Source	Sum of Squares	d.f.
Treatment	28	3
Error	20	4
Total	48	7

14.3 (a) We first find $k = 3$, $\bar{y}_1 = 5$, $\bar{y}_2 = 3$, $\bar{y}_3 = 1$, and $\bar{y} = 33/11 = 3$. Thus,

Obs.	Grand mean	Tr. Effect	Residuals
y_{ij}	$\bar{y}$	$\bar{y}_i - \bar{y}$	$y_{ij} - \bar{y}_i$

$$\begin{bmatrix} 7 & 5 & 4 & 4 \\ 6 & 1 & 2 \\ 2 & 1 & 0 & 1 \end{bmatrix} = \begin{bmatrix} 3 & 3 & 3 & 3 \\ 3 & 3 & 3 \\ 3 & 3 & 3 & 3 \end{bmatrix} + \begin{bmatrix} 2 & 2 & 2 & 2 \\ 0 & 0 & 0 \\ -2 & -2 & -2 & -2 \end{bmatrix} + \begin{bmatrix} 2 & 0 & -1 & -1 \\ 3 & -2 & -1 \\ 1 & 0 & -1 & 0 \end{bmatrix}$$

(b) Treatment SS $= 4(2^2) + 3(0^2) + 4(-2)^2 = 32$
    Residual SS $= 2^2 + 0^2 + \ldots + 0^2 = 22$
    Mean SS $\quad= 11(3)^2 = 99$
    Total SS $\quad\;= 7^2 + 5^2 + \ldots\; 0^2 + 1^2 = 153$
    Total SS (corrected) $= (7 - 3)^2 + (5 - 3)^2 + (4 - 3)^2 + (4 - 3)^2 + \ldots + (1 - 3)^2 =$
    $153 - 99 = 54$

(c) Treatment d.f. $= k - 1 = 3 - 1 = 2$
    Residual d.f. $\quad= \sum n_i - k = 4 + 3 + 4 - 3 = 8$
    Total d.f. $\qquad= 4 + 3 + 4 - 1 = 10$
(b) The analysis-of-variance table is

ANOVA Table

Source	Sum of Squares	d.f.
Treatment	32	2
Error	22	8
Total	54	10

14.5 We first find $\bar{y}_1 = 2$, $\bar{y}_2 = 3$, $\bar{y}_3 = 6$, $\bar{y}_4 = 4$, and $\bar{y} = \dfrac{48}{12} = 4$. Thus,

Obs.	Grand mean	Tr. Effect	Residuals
y_{ij}	$\bar{y}$	$\bar{y}_i - \bar{y}$	$y_{ij} - \bar{y}_i$

$$\begin{bmatrix} 2 & 1 & 3 \\ 1 & 5 \\ 9 & 5 & 6 & 4 \\ 3 & 4 & 5 \end{bmatrix} = \begin{bmatrix} 4 & 4 & 4 \\ 4 & 4 \\ 4 & 4 & 4 & 4 \\ 4 & 4 & 4 \end{bmatrix} + \begin{bmatrix} -2 & -2 & -2 \\ -1 & -1 & -1 \\ 2 & 2 & 2 & 2 \\ 0 & 0 & 0 \end{bmatrix} + \begin{bmatrix} 0 & -1 & 1 \\ -2 & 2 \\ 3 & -1 & 0 & -2 \\ -1 & 0 & 1 \end{bmatrix}$$

Treatment SS $\quad= 3(-2)^2 + 2(-1)^2 + 4(-2^2) + 3(0)^2 = 30$
Residual SS $\quad= 0^2 + (-1)^2 + 1^2 + \ldots + 0^2 + 1^2 = 26$
Total SS $\qquad= (2 - 4)^2 + (1 - 4)^2 + (3 - 4)^2 + (1 - 4)^2 + \ldots + (5 - 4)^2 = 56$
Treatment d.f. $\quad= k - 1 = 4 - 1 = 3$
Residual d.f. $\qquad= \sum n_i - k = 3 + 2 + 4 + 3 - 4 = 8$
Total d.f. $\qquad= 3 + 2 + 4 + 3 - 1 = 11$
The analysis-of-variance table is given by

ANOVA Table

Source	Sum of Squares	d.f.
Treatment	30	3
Error	26	8
Total	56	11

14.7 The overall mean is

$$\bar{y} = \frac{20 \times 94.4 + 20 \times 92.9 + 20 \times 75.5}{20 + 20 + 20} = \frac{5256}{60} = 87.6$$

Treatment SS $= 20\,(94.4 - 87.6)^2 + 20\,(92.9 - 87.6)^2 + 20\,(75.5 - 87.6)^2 = 4{,}414.80$

SSE $= 19\,(58.4)^2 + 19\,(54.2)^2 + 19\,(38.1)^2 = 148{,}196.39$

ANOVA Table

Source	Sum of Squares	d.f.
Treatment	4,414.80	2
Error	148,196.39	57
Total	152,611.9	59

14.9 (a) $F_{0.10}(3, 5) = 3.62$
   (b) $F_{0.10}(3, 10) = 2.73$
   (c) $F_{0.10}(3, 15) = 2.49$
   (d) $F_{0.10}(3, 30) = 2.28$
   (e) Increasing the denominator d.f. decreases the upper 10th percentile.

14.11 For $v_1 = 5$ and $v_2 = 30$ in the F-table, $F_{0.05}(5, 30) = 2.53$. We observe that

$$F = \frac{\text{Treatment } SS/(k-1)}{SSE/(n-k)} = \frac{23/5}{56/30} = 2.46$$

So, we fail to reject $H_0 : \mu_1 = \mu_2 = \mu_3 = \mu_4 = \mu_5 = \mu_6$, at level $\alpha = 0.05$.

14.13 We are to test the null hypothesis $H_0 : \mu_1 = \mu_2 = \mu_3$ versus the alternative hypothesis that the means are not all equal. Given $\alpha = 0.05$, the rejection region is determined by the value $F_{0.05}(2, 9) = 4.26$ obtained from the F-table. From Exercise 14.2, the observed value of F is

$$F = \frac{\text{Treatment } SS/(k-1)}{SSE/(n-k)} = \frac{312/2}{170/9} = 8.26$$

Consequently, we reject the null hypothesis that the means are equal at the $\alpha = 0.05$ level of significance.

14.15 We are to test the null hypothesis $H_0 : \mu_1 = \mu_2 = \mu_3$ versus the alternative hypothesis that the means are not all equal. Given $\alpha = 0.05$, the rejection region is determined by the value $F_{0.05}(2, 8) = 4.46$ obtained from the F-table. From Exercise 14.3, the observed value of F is

$$F = \frac{\text{Treatment } SS / (k-1)}{SSE / (n-k)} = \frac{32/2}{22/8} = 5.82$$

Consequently, we reject H_0 at the $\alpha = 0.05$ level.

14.17 For multiple-t confidence intervals we use $t_{\alpha/2m}$ with d.f. $= n - k$.

   (a) $\dfrac{\alpha}{2m} = \dfrac{0.05}{2(3)} = 0.00833$ and, with d.f. $= 26$, $t_{0.0083} = 2.559$

   (b) $\dfrac{\alpha}{2m} = \dfrac{0.05}{2(5)} = 0.005$ and, with df $= 26$, $t_{0.005} = 2.779$

14.19 The error d.f. $= (20 + 18 + 24 + 18 - 4) = 66$. The t intervals use $t_{0.025} = 2.00$ and the multiple-t intervals use $\dfrac{\alpha}{2m} = \dfrac{0.05}{2(6)} = 0.00417$ so we extrapolate $t_{0.00417} = 2.7$.

	(a) t-interval	(b) multiple-t interval
$\mu_1 - \mu_2$:	2.1 ± 2.08	2.1 ± 2.81
$\mu_1 - \mu_3$:	0.5 ± 1.94	0.5 ± 2.62
$\mu_1 - \mu_4$:	4.0 ± 2.68	4.0 ± 3.61
$\mu_2 - \mu_3$:	-1.6 ± 1.99	-1.6 ± 2.69
$\mu_2 - \mu_4$:	1.9 ± 2.71	1.9 ± 3.67
$\mu_3 - \mu_4$:	3.5 ± 2.60	3.5 ± 3.53

Observe that only μ_1 and μ_4 differ according to the multiple t-intervals.

14.21 The t-interval is given by $\overline{Y}_i - \overline{Y}_{i'} \pm t_{\alpha/2} S \sqrt{\dfrac{1}{n_i} + \dfrac{1}{n_{i'}}}$, while the multiple-$t$ interval is given by $\overline{Y}_i - \overline{Y}_{i'} \pm t_{\alpha/2m} S \sqrt{\dfrac{1}{n_i} + \dfrac{1}{n_{i'}}}$. So, the ratio of their lengths, namely,

$$\frac{2t_{\alpha/2} S \sqrt{\dfrac{1}{n_i} + \dfrac{1}{n_{i'}}}}{2t_{\alpha/2m} S \sqrt{\dfrac{1}{n_i} + \dfrac{1}{n_{i'}}}} = \frac{t_{\alpha/2}}{t_{\alpha/2m}}$$

does not depend on the data. For $m = 10$ and $\alpha = 0.10$, this ratio is

$$\frac{t_{0.05}}{t_{0.005}} = \frac{1.753}{2.947} = 0.595 \text{ for d.f.} = 15.$$

14.23 (a) We first find $k = 3$, $\overline{y}_{..} = 8$, $\overline{y}_{1.} = 7$, $\overline{y}_{2.} = 6$, $\overline{y}_{3.} = 11$. Also, $b = 4$, $\overline{y}_{.1} = 11$, $\overline{y}_{.2} = 8$, $\overline{y}_{.3} = 9$, and $\overline{y}_{.4} = 4$. Thus,

$$
\begin{array}{ccc}
\text{Obs.} & \text{Grand mean} & \text{Tr. Effect} \\
y_{ij} & \bar{y}_{..} & \bar{y}_{i.} - \bar{y}_{..}
\end{array}
$$

$$
\begin{bmatrix} 11 & 10 & 7 & 0 \\ 7 & 8 & 7 & 2 \\ 15 & 6 & 13 & 10 \end{bmatrix}
=
\begin{bmatrix} 8 & 8 & 8 & 8 \\ 8 & 8 & 8 & 8 \\ 8 & 8 & 8 & 8 \end{bmatrix}
+
\begin{bmatrix} -1 & -1 & -1 & -1 \\ -2 & -2 & -2 & -2 \\ 3 & 3 & 3 & 3 \end{bmatrix}
$$

$$
\begin{array}{cc}
\text{Bl. Effect} & \text{Error} \\
\bar{y}_{.j} - \bar{y}_{..} & y_{ij} - \bar{y}_{i.} - \bar{y}_{.j} + \bar{y}_{..}
\end{array}
$$

$$
+
\begin{bmatrix} 3 & 0 & 1 & -4 \\ 3 & 0 & 1 & -4 \\ 3 & 0 & 1 & -4 \end{bmatrix}
+
\begin{bmatrix} 1 & 3 & -1 & -3 \\ -2 & 2 & 0 & 0 \\ 1 & -5 & 1 & 3 \end{bmatrix}
$$

(b) The sums of squares are:

$$
\begin{array}{ll}
\text{Treatment SS} & = 4\,(-1)^2 + 4(-2)^2 + 4(3)^2 = 56 \\
\text{Block SS} & = 3(3)^2 + 3(0)^2 + 3(1)^2 + 3(-4)^2 = 78 \\
\text{Residual SS} & = 1^2 + 3^2 + (-1)^2 + \ldots + 1^2 + 3^2 = 64 \\
\text{Mean} & = 12(8)^2 = 768 \\
\text{Total SS} & = 11^2 + 10^2 + 7^2 + \ldots + 13^2 + 10^2 = 966 \\
\text{Total SS (corrected} & = 11^2 + 10^2 + 7^2 + \ldots + 13^2 + 10^2 - 12\,(8)^2 \\
& = 966 - 768 = 198
\end{array}
$$

(c)
$$
\begin{array}{ll}
\text{Treatment d.f.} & = k - 1 = 2 \\
\text{Block d.f.} & = b - 1 = 3 \\
\text{Residual d.f.} & = (k-1)\,(b-1) = (2)(3) = 6
\end{array}
$$

14.25 The null hypothesis is that the three treatment population means are the same; the alternative hypothesis is that they are not the same. The analysis-of-variance table is

Source of variation	Degrees of freedom	Sum of squares	Mean square	F
Treatments	2	56	28	2.62
Blocks	3	78	26	2.44
Error	6	64	10.667	
Total	11	198		

Since the critical value at the 0.05 level for an F distribution with 2 and 6 degrees of freedom is 5.14, we fail to reject the null hypothesis of equal treatment means. Since $F_{0.05}$ with 3 and 6 degrees of freedom is 4.76, the block effect is not significant.

14.27 (a) At each baking, select a loaf of bread from each recipe and randomize the position of the loaves in the oven.

(b) The Grand mean is $\overline{y}_{..} = \frac{1}{15}(0.95 + 0.71 + ... + 0.44) = \frac{10.77}{15} = 0.718$.

The Block means are 0.7833, 0.863, 0.6133, 0.7233, and 0.6067.
The recipe means are 0.796, 0.708, and 0.650.
$SS_B = 3[(0.7833 - 0.718)^2 = (0.7967 - 0.718)^2 + (0.6133 - 0.718)^2$
$\qquad + (0.7233 - 0.718)^2 + (0.6067 - 0.718)^2] = 0.2431$
$SS_T = 5[(0.796 - 0.718)^2 + (0.708 - 0.718)^2 + (0.610 - 0.718)^2] = 0.0540$
Total $SS = (0.95 - 0.718)^2 + ... + (0.44 - 0.718)^2 = 0.2971$
Hence, we have $SSE = 0.2971 - (0.0540 + 0.2431) = 0.0426$.
The ANOVA table is

Source of variation	Degrees of freedom	Sum of squares	Mean square	F
Treatments	0.0540	2	0.0433	8.17
Blocks	0.2431	4	0.0247	4.66
Residual	0.0426	8	0.0053	
Total	0.2971	14		

Since $F_{0.05}(2, 8) = 4.46 < 8.17$ we conclude that a significant treatment difference is indicated by the data. Also, $F_{0.05}(4, 8) = 3.84 < 4.66$, so the block effects are significant.

14.29 (a) The Grand mean is

$$\overline{y}_{..} = \frac{1}{36}(19.09 + 16.28 + ... + 21.58) = \frac{653.96}{36} = 18.1656.$$

The block means are 17.7533, 18.78, 18.2367, 17.94, 17.6867, and 18.5967.
The variety means are 19.6683, 17.1083, 17.2683, 17.7, 16.0767, and 21.177.
$SS_B = 6[(17.7533 - 18.1656)^2 + (18.78 - 18.1656)^2$
$\qquad + ... + (18.5967 - 18.1656)^2] = 6.112$
$SS_T = 6[(19.6683 - 18.1656)^2 + (17.1083 - 18.1656)^2$
$\qquad + ... + (21.1717 - 18.1656)^2] = 106.788$
$SS = (19.09 - 18.1656)^2 + (16.28 - 18.1656)^2$
$\qquad + ... + (21.58 - 18.1656)^2 = 117.898$

Hence, we have $SSE = 117.898 - (6.112 + 106.788) = 4.998$.
The ANOVA table is

Source of variation	Sum of squares	d.f.	Mean square	F-ratio
Treatments	106.788	5	21.358	106.79
Blocks	6.112	5	1.222	6.11
Error	4.998	25	0.200	
Total	117.898	35		

Since $F_{0.05}(5, 25) = 2.603 < 106.79$ we conclude that a highly significant treatment difference is indicated by the data. Also, $F_{0.05}(5, 25) < 6.11$ so the block effects are significant.

(b) Array of residuals

$$\begin{bmatrix} -0.166 & -0.416 & -0.546 & 0.212 & 0.586 & 0.331 \\ 0.007 & 0.157 & 0.287 & -0.264 & 0.229 & -0.416 \\ 0.571 & -0.299 & 0.041 & -0.181 & -0.268 & 0.137 \\ 0.157 & 0.687 & 0.487 & 0.166 & -1.071 & -0.426 \\ -0.569 & 0.091 & -0.449 & 0.159 & 0.372 & 0.397 \\ 0.001 & -0.219 & 0.181 & -0.091 & 0.151 & -0.023 \end{bmatrix}$$

The dot diagrams are given on the next page. Note that -1.071 is a possible outlier. The normal scores plot is also given on the next page (directly following the dot diagrams) and the same point is a little low.

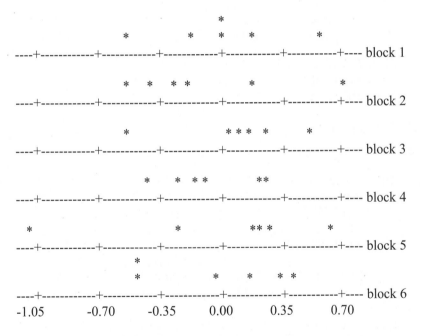

Dot diagrams for Exercise 14.29.

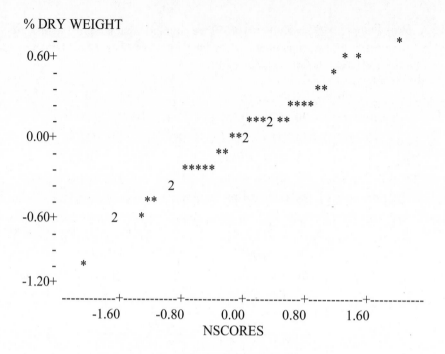

Normal-scores plot for Exercise 14.29.

14.31 Note that $k = 3$. We first find $\bar{y}_1 = 19, \bar{y}_2 = 13, \bar{y}_3 = 14, \bar{y} = 225/15 = 15$.

$$
\begin{matrix} y_{ij} \\ \begin{bmatrix} 19 & 18 & 21 & 18 \\ 16 & 11 & 13 & 14 & 11 \\ 13 & 16 & 18 & 11 & 15 & 11 \end{bmatrix} \end{matrix}
=
\begin{matrix} \bar{y} \\ \begin{bmatrix} 15 & 15 & 15 & 15 \\ 15 & 15 & 15 & 15 & 15 \\ 15 & 15 & 15 & 15 & 15 & 15 \end{bmatrix} \end{matrix}
$$

$$
\begin{matrix} \bar{y}_i - \bar{y} \\ + \begin{bmatrix} 4 & 4 & 4 & 4 \\ -2 & -2 & -2 & -2 & -2 \\ -1 & -1 & -1 & -1 & -1 & -1 \end{bmatrix} \end{matrix}
+
\begin{matrix} y_{ij} - \bar{y}_i \\ \begin{bmatrix} 0 & -1 & 2 & -1 \\ 3 & -2 & 0 & 1 & -2 \\ -1 & 2 & 4 & -3 & 1 & -3 \end{bmatrix} \end{matrix}
$$

14.33   (a) $F_{0.05}(7, 13) = 2.83$
        (b) $F_{0.05}(7, 20) = 2.51$
        (c) $F_{0.10}(7, 12) = 2.28$

14.35   The Minitab output is shown below:

```
One-way Analysis of Variance

Analysis of Variance
Source    DF      SS      MS      F       P
Factor     2    5.279   2.640  20.85   0.000
Error     14    1.772   0.127
Total     16    7.052
      Individual 95% CIs For Mean


Level    N    Mean    StDev   --+---------+---------+---------+-
Treat 1  5   1.4600  0.1306   (----*-----)
Treat 2  6   1.5967  0.4357    (-----*----)
Treat 3  6   2.6950  0.3886                        (----*----)
                              --+---------+---------+---------+-
Pooled StDev =   0.3558       1.20      1.80      2.40      3.00
                                        Moisture
```

14.37 (a) We first find $k = 3$, $\bar{y}_{..} = 8$, $\bar{y}_{1.} = 6$, $\bar{y}_{2.} = 7$, $\bar{y}_{3.} = 11$. Also, $b = 4$, $\bar{y}_{.1} = 7$,
$\bar{y}_{.2} = 12$, $\bar{y}_{.3} = 3$, and $\bar{y}_{.4} = 10$. Thus,

$$
\underset{\substack{\text{Obs.}\\ y_{ij}}}{\begin{bmatrix} 8 & 9 & 1 & 6 \\ 5 & 12 & 0 & 11 \\ 8 & 15 & 8 & 13 \end{bmatrix}} = \underset{\substack{\text{Grand mean}\\ \bar{y}_{..}}}{\begin{bmatrix} 8 & 8 & 8 & 8 \\ 8 & 8 & 8 & 8 \\ 8 & 8 & 8 & 8 \end{bmatrix}} + \underset{\substack{\text{Tr. Effect}\\ \bar{y}_{i.} - \bar{y}_{..}}}{\begin{bmatrix} -2 & -2 & -2 & -2 \\ -1 & -1 & -1 & -1 \\ 3 & 3 & 3 & 3 \end{bmatrix}}
$$

$$
+ \underset{\substack{\text{Bl. Effect}\\ \bar{y}_{.j} - \bar{y}_{..}}}{\begin{bmatrix} -1 & 4 & -5 & 2 \\ -1 & 4 & -5 & 2 \\ -1 & 4 & -5 & 2 \end{bmatrix}} + \underset{\substack{\text{Error}\\ y_{ij} - \bar{y}_{i.} - \bar{y}_{.j} + \bar{y}_{..}}}{\begin{bmatrix} 3 & -1 & 0 & -2 \\ -1 & 1 & -2 & 2 \\ -2 & 0 & 2 & 0 \end{bmatrix}}
$$

(b) The sums of squares are:

Treatment SS	$= 4(-2)^2 + 4(-1)^2 + 4(3)^2 = 56$
Block SS	$= 3(-1)^2 + 3(4)^2 + 3(-5)^2 + 3(2)^2 = 138$
Residual SS	$= 3^2 + (-1)^2 + \ldots + 2^2 + 0^2 = 32$
Total SS	$= 8^2 + 9^2 + 1^2 + \ldots + 8^2 + 13^2 - 12(8)^2 = 226$

(c) The $k = 3$ rows of the treatment array sum to zero, and the $b = 4$ columns of the block array sum to zero. All of the entries of the Error array sum to zero as do the columns and the rows. Consequently,

$$\text{Treatment d.f.} = k - 1 = 2$$
$$\text{Block d.f.} = b - 1 = 3$$
$$\text{Residual d.f.} = (k - 1)(b - 1) = (2)(3) = 6$$

Chapter 15

NONPARAMETRIC INFERENCE

15.1 (a) Rank collections for treatment B with sample sizes $n_A = 4$ and $n_B = 2$

Rank of B	Rank Sum W_B	Probability
1,2	3	1/15
1,3	4	1/15
1,4	5	1/15
1,5	6	1/15
1,6	7	1/15
2,3	5	1/15
2,4	6	1/15
2,5	7	1/15
2,6	8	1/15
3,4	7	1/15
3,5	8	1/15
3,6	9	1/15
4,5	9	1/15
4,6	10	1/15
5,6	11	1/15
	Total	1

When the two samples come from the same population, every pair of integers out of {1,2,3,4,5,6} is equally likely to be the ranks for the two B measurements. There are $\binom{6}{2} = 15$ potential pairs, so that each collection of possible ranks has a probability of 1/15.

(b) Both rank collections {1,4} and {2,3} have $W_B = 5$ so

$$P[W_B = 5] = 1/15 + 1/15 = 2/15.$$

233

Continuing, we obtain the distribution of W_B:

Values of W_B	3	4	5	6	7	8	9	10	11
Probability	$\frac{1}{15}$	$\frac{1}{15}$	$\frac{2}{15}$	$\frac{2}{15}$	$\frac{3}{15}$	$\frac{2}{15}$	$\frac{2}{15}$	$\frac{1}{15}$	$\frac{1}{15}$

These values agree with the tabulated entries of Table 7.

15.3 (a) Smaller sample size = 5, larger sample size = 6, so $P[W_S \geq 39] = 0.063$

   (b) Smaller sample size = 4, larger sample size = 6, so $P[W_S \leq 15] = 0.086$

   (c) Smaller sample size = 7, larger sample size = 7, $P[W_S \geq 66] = 0.049$, so c = 66.

15.5 (a)

Combined sample ordered observations	2.1	2.7	3.2	3.7	5.3
Ranks	1	2	3	4	5
Treatment	A	B	B	A	A

   We find $W_A = 1 + 4 + 5 = 10$.

   (b) Since n_B is the smaller sample size, $W_S = 2 + 3 = 5$.

15.7 We test H_0 : the populations are identical versus H_1 : the populations are different. Let W_S = rank sum of phosphate for the Chester White breed. The alternative is two-sided. From Appendix Table 9 we find with smaller size = 8 = larger size, $P[W_S \geq 87] = 0.025 = P[W_S \leq 49]$.

The combined ordered observations, with the Chester White underlined, are

Ordered observations	47	48	57	58	65	75	78	79
Ranks	1	2	3	4	5	6	7	8

Ordered observations	97	99	110	162	172	182	220	230
Ranks	9	10	11	12	13	14	15	16

We find $W_S = 1 + 2 + 3 + 5 + 6 + 9 + 10 + 11 = 47$

We conclude that the serum phosphate level is significantly different for the two breeds at the level $\alpha = 0.05$.

15.9 We test H_0: the populations are identical versus H_1: the populations are different. Let W_S = rank sum for treatment 2. The alternative is two-sided. From Appendix Table 9 we find with smaller size = 7 and larger size = 8, $P[WS \geq 73] = 0.027 = P[WS \leq 39]$. The sample sizes $n_A = 8$ and $n_B = 7$.

Combined ordered values	18	25	28	29	30	31	36	37	38	40	41	43	46	49	56
Ranks	1	2	3	4	5	6	7	8	9	10	11	12	13	14	15

$$W_S = 2 + 5 + 7 + 8 + 11 + 14 + 15 = 62$$

Consequently, we fail to reject the null hypothesis at level $\alpha = 0.054$.

15.11 (a) The configuration that most supports the alternative hypothesis is *BBBBBBBBBA* where the rank of the single *A* observation is 10.
   (b) There are 10 possible ranks (positions) for the single *A* and these are equally likely. Therefore, $P[W_A = 10] = 0.1$.
   (c) The single most extreme outcome has probability 0.1. An α of 0.05 cannot be achieved unless, whenever $W_A = 10$, we are willing to flip a coin to decide whether or not H_0 should be rejected.

15.13 The alternative is one-sided. We test H_0: there is no difference in treatments against H_1: $P[+] > 0.5$. Let S = number of positive signs among the differences. The rejection region is $R : S \geq c$. There are 13 positive values out of 18. From Appendix Table 2 with $n = 18$ we find $P[S \geq 13] = 0.048$. Since the observed value is $S = 13$, we reject H_0, at level $\alpha = 0.048$, in favor of $P[+] > 0.5$ or more than half of the population prefer recipe *A*.

15.15 The alternative is two-sided. We test H_0: no difference in treatments against H_1: $P[+] = 0.5$. Let S = number of positive signs among the differences. The rejection region is $R : S \leq c_1$ or $S \geq c_2$. Since there are 7 ties, the effective sample size is $25 - 7 = 18$. From Appendix Table 2 with $n = 18$, we find $P[S \leq 4] = 0.015 = P[S \geq 14]$. Since the observed value is $S = 11$, we conclude at $\alpha = 0.030$ that the data do not show a significant difference of opinion between husbands and wives.

15.17 (a) $P[T^+ \geq 54] = 0.034$,
   (b) $P[T^+ \leq 32] = 0.060$
   (c) Since $P[T^+ \geq 79] = 0.052$, we have $c = 79$.

15.19 (a) The alternative is two-sided. The rejection region is $R : T^+ \leq c_1$ or $T^+ \geq c_2$. From the table for the signed rank statistic, with $n = 6$ we find $P[T^+ \leq 0] = 0.016 = P[T^+ \geq 21]$. We calculate

Restaurant	1	2	3	4	5	6
Critic 1	6.1	5.2	8.9	7.4	4.3	9.7
Critic 2	7.3	5.5	9.1	7.0	5.1	9.8
difference $C2 - C1$	1.2	0.3	0.2	-0.4	0.8	0.1
sign	+	+	+	-	+	+
rank ordered $|C2 - C1|$	6	3	2	4	5	1

Since the observed value is $T^+ = 1 + 2 + 3 + 5 + 6 = 17$, we fail to reject H_0 at level $\alpha = 0.032$.

(b) We must consider extreme values in both tails. From the table with $n = 6$ we find $P[T^+ \leq 4] = 0.109 = P[T^+ \geq 17]$ so the observed value $T^+ = 17$ has significance probability $2(0.109) = 0.218$.

15.21 (a)

	Ranks 1, 2, 3	$T+$	Probability
	+, +, +	6	0.125
	+, +, -	3	0.125
	+, -, +	4	0.125
Signs	-, +, +	5	0.125
	+, -, -	1	0.125
	-, +, -	2	0.125
	-, -, +	3	0.125
	-, -, -	0	0.125

(b)

Values of T^+	0	1	2	3	4	5	6
Probability	0.125	0.125	0.125	0.250	0.125	0.125	0.125

From the table when $n = 3$, we see that the tail probabilities agree.

15.23 (a) The alternative is one-sided. We test H_0: there is no difference in treatments against H_1: $P[+] > 0.5$. There are two ties among the differences so the effective sample size is $n = 13$. Let $S =$ number of positive signs among the 13 differences *before – after*. The rejection region is $R : S > c$. From Appendix Table 2 with $n = 13$ we find $P[S \geq 10] = 0.046$.

(b) The alternative is one-sided so we reject for large values of T_+. Using the 0 differences for ranking and average rank for the other ties, we calculate

Ordered absolute value of differences	0	0	2	4	4	6	8	8
Ranks	1.5	1.5	3	4.5	4.5	6	7.5	7.5
Signs			+	+	+	+	+	+

Ordered absolute value of differences	10	10	10	18	18	26	32
Ranks	10	10	10	12.5	12.5	14	15
Signs	+	-	+	+	+	-	+

The observed value is $T^+ = 93$. There are ties, so the most accurate answer would be to find the distribution of T^+ under the tied structure (see Lehmann reference in text). To give a more expedient answer, we take a conservative approach and assign the difference -10 rank 11 and the two positive differences ranks 9 and 10. The resulting observed value of T^+ is then 92. If there were no ties, $R : T^+ \geq 990$ since $P[T^+ \geq 90] = 0.047$ for $n = 15$. Consequently, we reject H_0 and conclude that the mean blood pressure has been reduced.

15.25 We calculate

x	3.1	5.4	4.7
y	2.8	3.5	4.6
Ranks R_i	1	3	2
Ranks X_i	1	2	3

$$r_{S_p} = \frac{\sum_{i=1}^{3}\left(R_i - \frac{3+1}{2}\right)\left(S_i - \frac{3+1}{2}\right)}{\frac{3(3^2-1)}{12}}$$

$$= \frac{(1-2)(1-2)+(3-2)(2-2)+(2-2)(3-2)}{2} = 0.5$$

15.27 We are to test H_0 : independence against a two-sided alternative.

Student	1	2	3	4	5	6	7	8	9	10
Dexterity	23	29	45	36	49	41	30	15	42	38
Aggression	45	48	16	28	38	21	36	18	31	37
Ranks R_i	2	3	9	5	10	7	4	1	8	6
Ranks S_i	9	10	1	4	8	3	6	2	5	7

$$r_{S_P} = \frac{\sum_{i=1}^{10}\left(R_i - \frac{10+1}{2}\right)\left(S_i - \frac{10+1}{2}\right)}{\frac{10(10^2-1)}{12}}$$

$$= \frac{(2-5.5)(9-5.5)+(3-5.5)(10-5.5)+\cdots+(6-5.5)(7-5.5)}{82.5}$$

$$= -\frac{16.5}{82.5} = -0.20$$

Even though $n = 10$ is not large, we approximate that $\sqrt{n-1}\, r_{Sp}$ is approximately standard normal. Since the z value $\sqrt{9}\,(-0.20) = -0.60$ is not negative enough, we cannot reject the null hypothesis of independence.

15.29

Combined sample ordered observations	32	43	67	81	90	99
Ranks	1	2	3	4	5	6
Treatment	A	B	B	A	A	B

We find $W_A = 1 + 4 + 5 = 10$.

15.31(a) Rank collections for treatment A with sample sizes $n_A = 3$ and $n_B = 2$

Rank of A	Rank Sum W_A	Probability
1,2,3	6	1/10
1,2,4	7	1/10
1,2,5	8	1/10
1,3,4	8	1/10
1,3,5	9	1/10
1,4,5	10	1/10
2,3,4	9	1/10
2,3,5	10	1/10
2,4,5	11	1/10
3,4,5	12	1/10
Total		1

When the two samples come from the same population, every triple is equally likely to be the ranks for the three A measurements. There are $\binom{5}{3} = 10$ potential triples, so that each collection of possible ranks has a probability of 1/10.

(b) Both rank collections {1,2,5} and {1,3,4} have $W_A = 8$, so

$$P[W_A = 8] = 1/10 + 1/10 = 0.2.$$

Continuing, we obtain the distribution of W_A:

Values of W_A	6	7	8	9	10	11	12
Probability	0.1	0.1	0.2	0.2	0.2	0.1	0.1

15.33   (a) $P[T^+ \geq 28] = 0.098$,

      (b) $P[T^+ \leq 5] = 0.020$

      (c) Since $P[T^+ \leq 21] = 0.047$, we have $c = 21$.

15.35   We test H_0: populations A and B are identical versus H_1: they are different.

Combined ordered values	95	98	100	103	104	105	116	127	131	137
Ranks	1	2	3	4	5	6	7	8	9	10

Combined ordered values	140	149	150	151	155	164	167	178	179
Ranks	11	12	13	14	15	16	17	18	19

The rank sum of method 2 (the smaller sample) is

$$W_S = 1 + 2 + 3 + 4 + 6 + 7 + 8 + 9 + 11 = 51.$$

Referring to the table with smaller sample size = 9 and larger sample size = 10, we find that $P[W_S \geq 69] = 0.047 = P[W_S \geq 111]$. Since the observed value = 51 < 69, we conclude there is a significant difference at level $\alpha = 0.047 + 0.047 = 0.094$. In fact, the null hypothesis would be rejected even for α much smaller than 0.018.

15.37   (a)   For the Chester Whites,

Calcium	116	112	82	63	117	69	79	87
Phosphate	47	48	57	75	65	99	97	110
Ranks R_i	7	6	4	1	8	2	3	5
Ranks S_i	1	2	3	5	4	7	6	8

$$r_{S_P} = \frac{\sum_{i=1}^{3}(R_i - 4.5)(S_i - 4.5)}{\dfrac{8(8^2 - 1)}{12}} = \frac{-22}{42} = -0.524$$

  (b)   We test the null hypothesis H_0 : independence against a two-sided alternative. The value of the test statistic is $\sqrt{n-1}(-0.524) = -1.386$. Consequently, we fail to reject the hypothesis of independence.

(c)   If we approximate that $\sqrt{n-1}\,r_{S_P}$ is nearly standard normal, the level of significance is
$$\alpha = P[Z < -1.96] + P[Z > 1.96] = 0.025 + 0.025 = 0.05$$
However, $n = 7$ may not be large enough for a good approximation.

15.39   (a)   We calculate

x	10	7	8
y	15	13	9
difference $d = y - x$	5	6	1
sign	+	+	+
Ranks absolute d	2	3	1

$$S = 1 + 1 + 1 = 3$$

(b) $T^+ = 1 + 2 + 3 = 6$

✍ **Notes!**

✍ Notes!

✍ Notes!

✍ Notes!

✍ Notes!

✍ Notes!

✍ **Notes!**

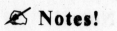 **Notes!**

✍ Notes!

✍ Notes!

✍ **Notes!**

✍ **Notes!**

✍ Notes!

✍ Notes!

✍ Notes!

✍ Notes!

✍ Notes!

✍ Notes!

✍ Notes!

✍ Notes!